Manual on Significance of Tests for Petroleum Products: 5th Edition

GEORGE V. DYROFF
editor

ASTM Manual Series: **MNL 1**
Revision of Special Technical Publication (STP) 7C

Library of Congress Cataloging-in-Publication Data

Manual on significance of tests for petroleum products: 5th edition/ George V. Dyroff, editor.
(ASTM manual series; MNL 1)
"ASTM publication code number (PCN) 28-001089-12."
Rev. ed. of: Significance of tests for petroleum products/ ASTM Committee D-2 on Petroleum Products and Lubricants, 1977.

ISBN 0-8031-1207-6
1. Petroleum—Testing. 2. Petroleum products—Testing.
I. Dyroff, George V. II. ASTM Committee D-2 on Petroleum Products and Lubricants. Significance of tests for petroleum products. III. Series.
TP691.M36 1989 88-36699
CIP

ASTM Publication Code Number (PCN) 28-001089-12
ISBN 0-8031-1207-6

Library of Congress Catalog Card Number: 76-29513

NOTE: The Society is not responsible, as a body, for the statements and opinions advanced in this publication.

Printed in Baltimore, MD
February 1989

Foreword

THIS ASTM MANUAL HAS BEEN DEVELOPED through the joint efforts of the American Society for Testing and Materials (ASTM) and the Institute of Petroleum (IP) and is the fifth edition of ASTM Special Technical Publication (STP) 7C. ASTM, particularly its Committee D-2 on Petroleum Products and Lubricants, and IP are long-established sources of standard testing and evaluation methods for petroleum and petroleum products. Many of these standard methods are accepted internationally as yardsticks for the determination of product quality.

The purpose of this volume is to provide an informative reference text on the significance of ASTM and IP test results that are used in commercial transactions to definitively describe petroleum raw materials and petroleum products. To ensure orderly treatment, separate chapters have been prepared for each major raw material or end product. The general approach in each chapter is as follows:

(a) A definitive description of the raw material or product is presented along with its known or prospective use applications.

(b) The primary quality characteristics that define the material in those use applications are outlined.

(c) The particular tests and analytical procedures applicable to the quality assessment for the material are indicated.

(d) Suggestions for further reading are listed.

Each chapter has been written as an entity. Since many tests are applicable to several products, this results in a certain amount of repetition. Repetition was believed to be preferable to elaborate cross referencing.

Although this book will be revised periodically, there will be no assurance that the test methods or specifications cited are current. Up-to-date standards can be found in the appropriate volumes of the *Annual Book of ASTM Standards* and *IP Standards for Petroleum and Its Products.*

In this revision, Chapter 4 on Automotive Gasoline has been completely redone. Chapter 5 on Aviation Fuels and Chapter 6 on Diesel Engine and Nonaviation Gas Turbine Fuels have been extensively revised. In addition, the entire book has been updated.

Acknowledgments

THIS VOLUME HAS BEEN DEVELOPED over a number of years through the joint efforts of the American Society for Testing and Materials and the Institute of Petroleum. In the process, chapters have been written, revised, and rewritten until it became difficult to distinguish between author, reviewers, and editors. Hopefully, the following list of acknowledgments includes all who contributed to the current edition.

Editor

George V. Dyroff

Crude Oils

J. O'Donnell	BP Research Centre, U.K.
C. E. Webber	Sun Oil Co.

Gaseous Fuels and Light Hydrocarbons

R. E. Cannon	Gas Processors Association
C. A. Munson	Petro-Tex Chemical Corp.
F. W. Selim	Phillips Petroleum Co.

Petroleum Solvents

D. M. Fenton	Union Oil Co. of California
A. J. Goodfellow	Carless, Capel and Leonard Ltd., U.K.
J. F. Hickerson	Exxon Co.
S. A. Yuhas	Exxon Chemicals

Automotive Gasoline

C. H. Ruof	
J. H. Macpherson	Chevron Research Co.

Aviation Fuels

G. J. Bishop	Shell International Petroleum Co., Ltd., U.K.

Diesel Engine and Nonaviation Gas Turbine Fuels

H. E. Doering	
C. H. Jones	Midstream Fuel Services, Inc.
E. W. White	U.S. Naval Ship Research and Development Center
G. K. Brower	International Harvester Co.
W. H. Kite, Jr.	Exxon Research and Engineering Co.
J. A. McLain	Caterpillar Tractor Co.
R. E. Pegg	Esso Research Centre, U.K.
C. C. Ward	U.S. Department of the Interior, Bureau of Mines

Heating and Power Generation Fuels

J. R. Callaway	Texaco, Inc.
W. H. Kite, Jr.	Exxon Research and Engineering Co.
N. J. Schlesser	Atlantic Richfield Co.
G. G. Stephens	BP Research Centre, U.K.

Lubricating Oils

J. B. Berkley	Mobil Oil. Co., Ltd., U.K.
J. W. Gaynor	Amoco Oil Co.

Lubricating Greases
R. S. Barnes — Amoco Chemicals Corp.
E. A. Goodchild — Hoffman Manufacturing Co., Ltd., U.K.
A. T. Polishuk — Sun Oil Co.
Petroleum Waxes, Including Petrolatums
J. J. Kaufman — Witco Corp.
White Mineral Oils
J. J. Kaufman — Witco Corp.
Robert Simonoff — Witco Corp., Sonneborn Division
V. Biske — Burmah Oil Trading, Ltd., U.K.
C. F. W. Gebelein — Pennsylvania Refining Co.
Petroleum Oils for Rubber
J. S. Sweely — Sun Oil Co.

Contents

	Introduction	**1**
1	**Crude Oils**	**3**
	Introduction	3
	Sampling	3
	Preliminary Evaluation or Assay	3
	Full or Comprehensive Evaluation	6
	Crude Oil as a Power Generation Fuel	9
	Applicable ASTM/IP Standards	9
	Bibliography	10
2	**Gaseous Fuels and Light Hydrocarbons**	**11**
	Introduction	11
	Quality Criteria	13
	Applicable ASTM Specification	15
	Applicable ASTM/IP Standards	15
	Bibliography	17
3	**Petroleum Solvents**	**18**
	Introduction	18
	General Uses	18
	Product Requirements	18
	Types of Hydrocarbons	20
	Commercial Hydrocarbon Solvents	22
	Test Methods for Solvents	24
	Applicable ASTM Specifications	27
	Applicable ASTM/IP Standards	27
	Bibliography	30
4	**Automotive Gasoline**	**31**
	Introduction	31
	Grades of Gasoline	31
	Antiknock Performance	32
	Volatility	33
	Other Properties	35
	U.S. Legal Restrictions on Blending Additives and Alternative Fuels into Gasoline	38
	Gasoline-Oxygenate Blends	39
	Applicable ASTM Specifications	42
	Applicable ASTM/IP Standards	42

5 Aviation Fuels **45**
Introduction 45
Historical Development of Aviation Fuels 45
Aviation Gasoline 45
Aviation Turbine Fuels (Jet Fuels) 52
Aviation Fuel Additives 65
Additive Tests 69
Applicable ASTM Specifications 70
Applicable ASTM/IP Standards 70
Bibliography 73
Appendix I 73

6 Diesel Engine and Nonaviation Gas Turbine Fuels **74**
Introduction 74
Combustion Process 75
General Fuel Characteristics and Specifications 78
Fuel Properties and Tests 81
Heat of Combustion 87
Applicable ASTM Specifications 93
Applicable ASTM/IP Standards 93
Bibliography 94

7 Heating and Power Generation Fuels **95**
Introduction 95
Burning Equipment 98
Fuel Oil Classification and Specifications 101
Fuel Oil Laboratory Tests and Their Significance 101
Domestic Heating Oil Performance Evaluation 109
Kerosine Performance Evaluation 109
Applicable ASTM Specification 111
Applicable ASTM/IP Standards 112
Bibliography 113

8 Lubricating Oils **114**
Introduction 114
Composition and Manufacture 115
General Properties 116
Automotive Engine Oils 119
Marine Diesel Engine Oils 121
Industrial and Railway Engine Oils 122
Gas-Turbine Lubricants 123
Gas Engine Oils 124
Gear Oils 125
Automatic Transmission Fluids 126
Steam-Turbine Oils 127
Hydraulic Oils 128
Other Lubricating Oils 129
Applicable ASTM/IP Standards 130

9 Lubricating Greases **132**
Introduction 132

Composition 132
Properties 133
Evaluation of Properties and Significance 133
Applicable ASTM/IP Standards 141

10 Petroleum Waxes, Including Petrolatums 143
Introduction 143
Occurrence and Refining of Waxes 143
Definitions 144
Applications for Wax 144
Quality Criteria 145
Test Methods 150
Inspections of Typical Petroleum Waxes 155
Applicable ASTM/IP Standards 155
Bibliography 156

11 White Mineral Oils 157
Introduction 157
Manufacture 157
Purity Guardianship 157
Assessment of Quality 158
Applicable ASTM/IP Standards 162

12 Petroleum Oils for Rubber 164
Introduction 164
Composition of Rubber Oils 165
Importance of Composition to Compounders 166
Classification of Rubber Oils 166
Physical Properties of Rubber Oils 167
Specialty Application of Process Oils 169
Applicable ASTM/IP Standards 169

Introduction

PREPARING A RELIABLE REFERENCE GUIDE on the significance of tests for petroleum and petroleum products is no simple matter. Test methods define product quality. Quality, in turn, can be defined as the "fitness" of the material for its intended use. The history of petroleum is a story of a wide array of end products with continually changing quality criteria.

For those interested in product quality and its assessment, this means the situation is never static. There are no frozen standards in the petroleum and natural gas industries. Many well-established test procedures, which at one time may have afforded an adequate quality assessment of raw materials or end products or both, have become obsolete and produce results of limited significance. A number of these old time testing procedures have been replaced by radically different tests and analytical procedures that provide more accurate and meaningful appraisal of the qualities of materials. Other prospective new methods are being examined or under development.

The ultimate adoption of any of these new testing or evaluation methods will not constitute a "final chapter" on the subject of quality criteria and assessment. The state of the art is in a continual state of flux. The need for still another chapter is always "just around the corner."

1

Crude Oils

INTRODUCTION

WHILE CRUDE OIL OCCURS UNDERGROUND in most of the world, its origin continues to be the subject of speculation by scientists. Crude oils are organic in nature and made up of combinations of carbon and hydrogen with lesser amounts of sulfur, nitrogen, oxygen, and traces of vanadium, nickel, and other metals. The compounds of carbon and hydrogen are complex but are principally paraffinic, naphthenic, and aromatic. Any given crude oil consists of thousands of compounds, its complexity increasing with the boiling range of the oil.

There is no overall or generally accepted method of classifying crude oils. What is termed an ultimate analysis gives the composition as a percentage of the principal elements carbon, hydrogen, nitrogen, oxygen, and sulfur. A chemical analysis gives the composition in percentages of paraffinic, naphthenic, and aromatic type compounds. These types of analyses are valuable in formulating a general idea of the usefulness of the crude in producing various refinery products, but the analyses give little indication of the quantity of the various products that can be produced. To obtain this information, additional evaluation is necessary.

SAMPLING

Since many of the properties of crude oils change with change in temperature and pressure or lengthy exposure to atmospheric environment, proper sampling is very important. If the dissolved light hydrocarbons such as methane, ethane, propane, and butane are permitted to escape by exposure to the atmosphere, qualities such as gravity, vapor pressure, octane number, and distillation tests will change. The basic principle of sampling is to obtain a sample or a composite of several samples in such a manner and from such locations in the oil field separator, storage tank, pipeline, or other container that the sample will be truly representative of the crude oil or product.

Sample containers must be of a material that will not react with any component of the crude oil; for example, hydrogen sulfide will react with ordinary steel containers. Sampling may be individual or continuous, and different sampling procedures are sometimes required for different laboratory tests.

Problems of sampling and procedural requirements are presented in ASTM Practice for Manual Sampling of Petroleum and Petroleum Products (D 4057).

PRELIMINARY EVALUATION OR ASSAY

The preliminary crude oil assay provides general data on the oil and is based upon a number of standard tests. The tests consist primarily of a fractional distillation of the oil, a hydrocarbon analysis showing the volume percent of components lighter than hexane, and a series of tests used by the refiner to determine how the crude will yield the desired products from a given operation. Some of the more important tests are used to determine gravity, sulfur content, salt content, water and sediment, viscosity, pour point, and metallic contaminants. These tests, and fractional distillation, hydrocarbon analysis, and characterization factors, will be discussed briefly, and test method references will be furnished. (Table 1 presents preliminary evaluations on

TABLE 1. Preliminary evaluation of typical crude oils.

	Texas Gulf Coast Mix	Venezuela (Lagomar)	Nigerian (Bonney)	Canada (Redwater)
PROPERTIES				
Gravity (API at 15.5°C or 60°F)	36.5	30.7	38.1	34.9
Paraffin wax, weight %	1.84	...	...	...
Sulfur, weight %	0.16	1.48	0.14	0.56
Salt, lb/1000 bbl	7.0	<2.0	<2.0	3.0
Viscosity, SUS at 25°C (77°F)	41.0	107.0	38.4	47.8
Viscosity, SUS at 38°C (100°F)	37.3	64.3	35.7	41.7
Characterization factor (midboiling point)	11.76	11.95	11.90	11.89
Flash point, °C (°F)	< −4(40)	< −4(40)	< −4(40)	< −4(40)
Pour point, °C (°F)	−15(5)	−12(10)	2(35)	−12(10)
Water and sediment, volume %	0.1	trace	trace	trace
Color	2820	30160	2950	19340
Reid vapor pressure at 38°C (100°F)	3.2	2.5	6.9	8.4
Total acid number (Neutral Number)	0.56	0.11	0.24	0.23
Hydrogen sulfide, lb/1000 bbl	<0.5	0.8	0.0009	87.2
Mercaptans, lb/1000 bbl	<0.5	0.3	<0.0001	16.1
Iron, ppm	8.0	10.0	7.0	5.0
Nickel, ppm	0.9	17.0	5.0	12.0
Vanadium, ppm	0.7	175.0	0.3	3.0
Copper	0	0	0	0
DISTILLATION				
Initial boiling point: °C (°F)	< 10(50)	< 10(50)	< 10(50)	< 10(50)
5%	91(196)	95(203)	81(178)	68(155)
10%	118(245)	132(269)	104(220)	110(230)
20%	161(322)	187(369)	143(289)	148(299)
30%	203(398)	217(422)	191(375)	194(382)
40%	238(461)	297(566)	232(450)	247(477)
50%	267(513)	358(677)	270(518)	301(573)
60%	300(572)	420(788)	311(592)	360(680)
70%	346(654)	486(907)	362(684)	422(792)
80%	404(759)	564(1047)	416(780)	478(892)
90%	473(884)	... ...	481(897)	556(1032)
95%	538(1000)	... ...	534(994)	... ...
Endpoint, °C (°F)	556(1063)	584(1083)	563(1045)	566(1050)
Recovery, %	96	84	96	92
Residue, %	4	16	4	8
Loss	0	0	0	0
LIGHT HYDROCARBON ANALYSIS VOLUME %				
Ethane	0.04	0.07	0.11	0.16
Propane	0.28	0.38	0.95	1.13
I-butane	0.32	0.24	0.92	0.68
Butane	0.71	0.91	1.96	2.23
I-pentane	0.91	0.74	1.69	1.65
Pentane	0.97	1.29	1.57	1.93
2-methyl pentane	0.82	0.77	1.02	1.18
3-methyl pentane	0.40	0.37	0.53	0.75
Hexane and heavier	95.55	95.23	91.25	90.28

four typical crude oils from the United States, Venezuela, Nigeria, and Canada).

Gravity

The American Petroleum Institute (API) gravity of a crude enters into most pricing structures, and it is used for the correction of temperature and volume in bulk oil measurement. The test for API gravity is a simple hydrometer method which is described in ASTM Test for API Gravity of Crude Petroleum and Petroleum Products (Hydrometer Method) (D 287).

Sulfur, Hydrogen Sulfide, and Mercaptan Sulfur

The sulfur content of crude oil is an important quality because the complexity and expense of the refining operation increase tremendously as the sulfur content of the crude increases. Sulfur content varies from practically nothing to as high as several percent in some crudes.

Hydrogen sulfide is evolved during the distillation process. Generally, the other sulfur compounds concentrate in the heavier distillation residue. Residual fuel oils containing more than very small amounts of sulfur are subjected to severe price penalties at the marketplace. In addition, they are often barred as fuel by federal, state, or local environmental laws.

Sulfur in crude oils is usually determined by oxidizing a sample in a bomb and converting the sulfur compounds to barium sulfate which is determined gravimetrically, as described in ASTM Test for Sulfur in Petroleum Products (General Bomb Method) (D 129/IP 61).

Hydrogen sulfide and mercaptan sulfur can be estimated by several methods. Refineries choose the one best suited to their situation. One estimation of hydrogen sulfide can be made by distilling out the dissolved gas, absorbing it in a suitable chemical solution, and analyzing the solution. A more frequently used method for determining hydrogen sulfide and mercaptan sulfur is by a potentiometric titration in which silver sulfide and silver mercaptides are determined in a nonaqueous solution.

Salt Content

The salt content of crudes varies over a rather wide range, which results principally from two factors—production practices in the oil fields and the handling of the crude in tankers bringing it to the refinery. Even a very small salt content is critical to refinery operations because of the constant accumulation of salt in key units of the refinery such as stills, heaters, and exchangers. In addition to caking up this equipment, certain metallic salts break down and liberate corrosive acids during the processing operation.

A widely employed test for inorganic chlorides is by potentiometric titration in a nonaqueous solution as described in ASTM Test for Salt in Crude Oil (Electrometric Method) (D 3230). Refiners normally use adaptations of this type of test. Settling, heating, chemical treating, and mixing with water all reduce salt content. In stubborn cases, a high potential electric field across the settling vessel is helpful.

Water and Sediment

In refining operations, the presence of water and sediment leads to major difficulties such as corrosion, uneven heating and plugging in heaters and exchangers, and adverse effects on product quality.

Sediment normally exists in crude oils as extremely fine, well-dispersed solids. Solids that originate in the reservoir from which the crude came, or in drilling fluids used to drill the wells, may take the form of sand, clay, shale, or rock particles. Other sediments such as scale can be picked up from tubing, pipe, tanks, and other production and transportation equipment. Water may appear in the crude as droplets or as emulsion and can contain chemical salts and other harmful substances.

Methods used for the determination of water and sediment in the oil fields differ from those used in the refinery for a number of practical and economic reasons. The field operator selects one of three centrifugal test methods best suited to the crude oil and the particular operating situation. Usually waxy crude samples are heated. An aromatic solvent such as toluene is used with asphaltic crudes, and other samples are treated with emulsifiers. In the refinery, the water is measured by distillation and the sediment by extraction. These procedures are described in ASTM Tests for Water and Sediment in Crude Oils (D 96), ASTM Test for Water in Petroleum Products and Bituminous Materials by Distillation (D 95/IP 74), and ASTM Test for Sediment in Crude and Fuel Oils by Extraction (D 473/IP 53).

Viscosity and Pour Point

Although the viscosity and viscosity variation with change in temperature are important for establishing performance standards of lubricating oils, viscosity determinations on crude oils are used principally in the design of field gathering systems and the lines and pumps between refinery storage and the processing facility. In these systems, viscosity and pour-point determinations provide the keys for solving transportation problems associated with crude oils.

Viscosity is determined at two different temperatures—25°C (77°F) and 100°C (212°F). The procedure for determining kinematic viscosity is preferred today. However, the results of the test are often converted to Saybolt units for reporting purposes. The kinematic viscosity is determined by measuring the time for the liquid to flow under gravity through a glass viscometer, and it is described in ASTM Determination of Kinematic Viscosity of Transparent and Opaque Liquids (and the Calculation of Dynamic Viscosity) (D 445/IP 71). Tables for converting to Saybolt units appear in ASTM Conversion of Kinematic Viscosity to Saybolt Universal Viscosity or to Saybolt Furol Viscosity (D 2161).

Metallic Contaminants

Metallic trace components such as iron, sodium, nickel, vanadium, lead, and arsenic have an adverse effect on refining or processing operations. These materials are catalyst poisons, and some of them do additional harm. For example, vanadium, compounds damage turbine blades and refractory furnaces. Sodium can cause problems in furnace brickwork.

Since numerous laboratory techniques are available to analyze a crude oil for metallic contaminants, the one used is a matter of individual preference. However, one such procedure is described in ASTM Test for Trace Metals in Gas Turbine Fuels (Atomic Absorption Method) (D 2788).

Distillation

In the preliminary assay of crude oils, a distillation is performed under poor conditions of fractionation. Temperatures of the distilled vapor are recorded at initial boiling points and at the following volume percentage fractions of the charge in the receiver: 5, 10, 20, 30, 40, 50, 60, 70, 80, 90, 95, and at the end point. Crudes will crack at temperatures above 260 to 316°C (500 to 600°F), as evidenced by an increase in the distilling rate while the vapor temperature stabilizes and the appearances of smokey vapor. For this reason, many assay laboratories interrupt the distillation at a temperature around 232°C (450°F) and then reduce the pressure to 1 mm Hg. This permits the distillation to continue at reduced temperatures right up to the end point with no danger of cracking. The temperature readings are converted from actual temperatures at 1 mm pressure to the corresponding temperatures at atmospheric pressure.

Apparatus and procedure for this type distillation are described in ASTM Test for Distillation of Petroleum Products at Reduced Pressure (D 1160). If desired, this procedure can be easily modified by starting the distillation at atmospheric pressure and finishing under vacuum.

Light Hydrocarbon Analysis

Most crude oil assays and all comprehensive or full examinations include an analysis of the light hydrocarbons dissolved in the oil, usually ethane through pentane or sometimes hexane. The apparatus used is referred to frequently as a gas liquid partition chromatograph. Any chromatograph having a thermostated oven and detection system of proper sensitivity is satisfactory. A description of the test is found in ASTM Determination of C_2 through C_5 Hydrocarbons in Gasolines by Gas Chromatography (D 2427).

Characterization Factor

A tool used by refiners to correlate some of the important characteristics of crude oil is an empirical relationship generally referred to as the Characterization Factor. There are variations or modifications of the original concept developed by Watson, Nelson, and Murphy[1] but basically it follows

$$K = \sqrt[3]{\frac{T_B}{S}}$$

where

T_B = molal average boiling point in Rankine temperature, and
S = specific gravity at 15.5°C (60°F).

FULL OR COMPREHENSIVE EVALUATION

The tests utilized in making a preliminary assay are relatively simple. The results give a useful and general picture of the quality of the oil, but they do not provide

[1]Watson, Nelson, and Murphy, *Industrial and Engineering Chemistry*, Vol. 27, 1935, p. 1460.

sufficient data for an economic evaluation or adequate knowledge of the quality and quantity of the products the crude can yield. A more comprehensive evaluation is required to supply this information, and each refiner tailors the full assay to meet his particular needs—good lube stock, maximum motor fuel production, jet and diesel fuels, etc. The evaluation starts with a true boiling point distillation. Simply stated, this is a distillation of the crude oil under good fractionating conditions. The equipment consists of a kettle, heater, packed fractionating column, condenser, receiver, temperature recorders, and the necessary vacuum pump and control apparatus to permit the later stages of the distillation to be conducted under vacuum. Sometimes a secondary condenser, using a solid carbon dioxide and acetone or other suitable liquid coolant, will be added for recovering the lighter boiling fractions such as propane and butanes. This also can include apparatus for recovering and measuring the amount of vented hydrogen sulfide gas by an agent such as acidified cadmium chloride solution.

Samples of overhead distillate between desired temperature intervals are removed from the distillate receiver, measured or weighed, and then analyzed individually or in blends to appraise their value as motor fuel, fuel oil, reformer charge, etc. The true boiling point curve is plotted showing volume percent distilled versus temperature. Figure 1 is such a curve for a typical crude oil. Looking at Fig. 1, it can be seen that liquefied petro-

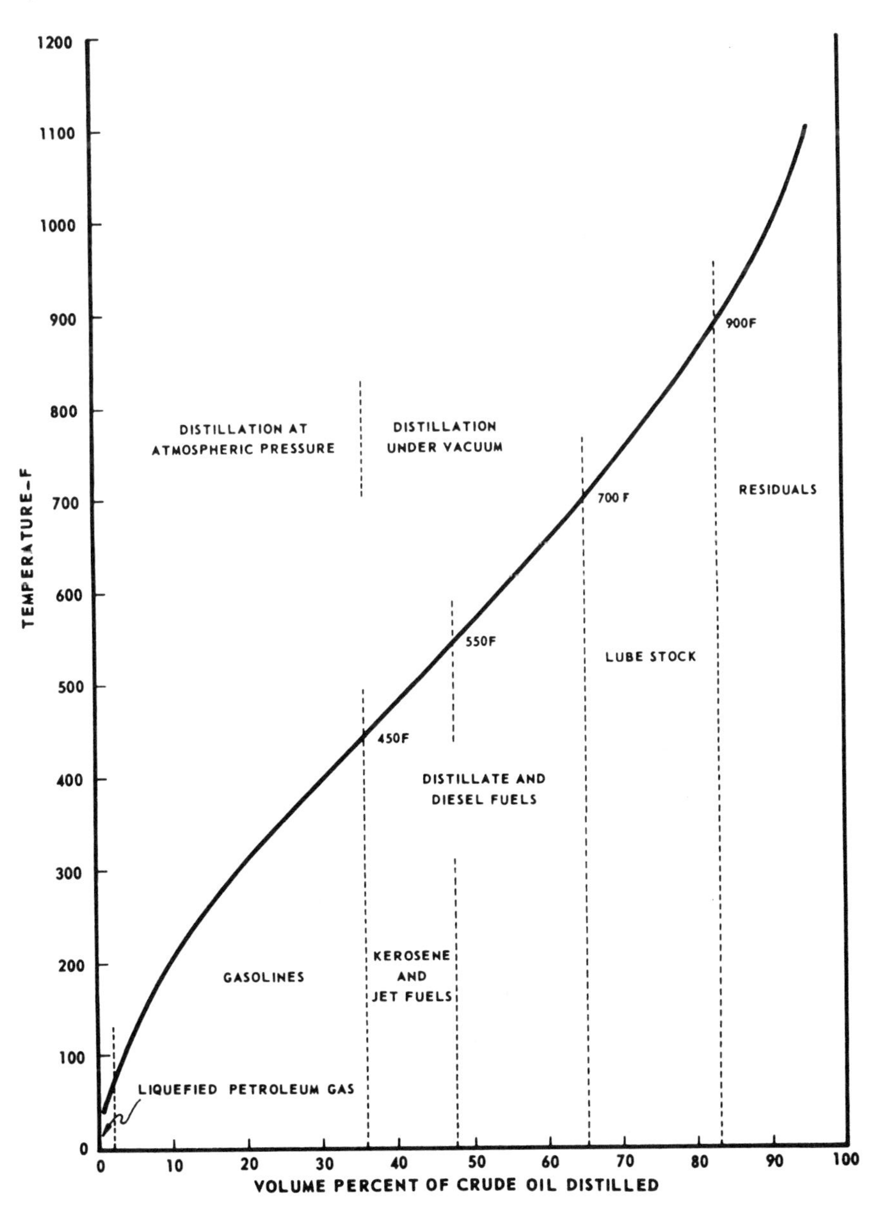

FIG. 1—True boiling point distillation curve Louisiana-Mississippi crude mix.

leum gas, constituting 1.6 percent of the sample, distills at temperatures below 15.6°C (60°F), gasoline distills at temperatures up to 282°C (540°F), kerosine and jet fuels up to 288°C (550°F), distillate and diesel fuels are in the 232 to 371°C (450 to 700°F) range and so on until residual fuel oil and asphalt comprise the residue at temperatures above 482°C (900°F). The procedure for conducting a true boiling point distillation is described in ASTM Distillation of Crude Petroleum (15 Theoretical Plate Column) (D 2892).

A few of the principal properties taken

TABLE 2. Portions of a comprehensive evaluation of crude oil Louisiana-Mississippi crude oil mix.

PROPERTIES	
Gravity, API at 15°C (60°F)	38.3
Specific gravity, 15°C (60°F)	0.8333
Paraffin wax, weight %	4.71
Sulfur, weight %	0.10
Salt, lb/1000 bbl	15.0
Viscosity, SUS at 25°C (77°F)	47.2
Viscosity, SUS at 37°C (100°F)	40.4
Characterization factor (midboiling point)	12.09
Flash point, °C (°F)	3.8(39.0)
Pour point, °C (°F)	−15.0(5.0)
Water and sediment, volume %	0.04
Color	5220.0
Reid vapor pressure at 37°C (100°F)	3.2
Total acid number (neutral number)	0.18
Hydrogen sulfide lb/1000 bbl	0.4
Mercaptans, lb/1000 bbl	0.4
DISTILLATION	
Initial boiling point, °C (°F)	9(49.0)
5%	65(149.0)
10%	97(208.0)
20%	147(298.0)
30%	202(397.0)
40%	251(484.0)
50%	294(562.0)
60%	342(648.0)
70%	398(749.0)
80%	460(860.0)
90%	533(992.0)
95%	. . .
End point	576(1070.0)
Residue, %	6.0
Loss	0
LIGHT HYDROCARBON ANALYSIS VOLUME %	
Ethane	0.02
Propane	0.34
I-butane	0.42
Butane	0.84
I-pentane	1.21
Pentane	1.15
2-methyl pentane	1.13
3-methyl pentane	0.57
Hexane and heavier	94.32
PRODUCT EVALUATION	
Butanes and lighter, volume %	1.6
Debutanized straight run gasoline, volume %	3.2
Octane number clear (BON)	75.3
Motor reformer charge, volume %	11.3
Fresh heavy naphtha, volume %	7.8
Heavy gas oil, volume %	47.7
Vacuum tower distillate, volume %	7.2
Vacuum tower bottoms, volume %	9.5
Potential benzene	0.44
Potential toluene	2.00
Potential xylene	1.24

from a comprehensive evaluation of a Louisiana-Mississippi crude oil mix are shown in Table 2. Space does not permit inclusion of all the various test results on individual distillate cuts necessary to evaluate them for lube stock, reformer charge, solvent refining potential, motor fuel, and others.

CRUDE OIL AS A POWER GENERATION FUEL

The use of crude oil as a power generation fuel recently has received considerable attention. The factors that have promoted this use are fuel availability, cost, and environmental aspects. The energy shortage has prompted electric utilities to resort to the burning of crude oil under the boilers of power stations and to use crude oil as the fuel for gas-turbine operations of standby generators.

The specifications for crude oils for this purpose have not been established, but there are known critical factors that must be considered. Since crude oils normally have relatively low-flash points, their use has associated fire hazards and explosive aspects. This fact necessitates the use of proper storage and handling facilities. These facilities must be explosion proof, vapor proof, and vapor retaining. Furthermore, the crude oil must be fluid or pumpable, and it must have low ash and sulfur contents to meet emission standards that satisfy environmental regulations. If these properties are not prevalent in the crude oil, sulfur dioxide and fly ash scrubbing or removal equipment must be provided to assure compliance.

Crude oils used as gas-turbine fuel also must have very low-metallic contents to ensure adequate turbine blade performance and comply with the manufacturer's requirements.

Standard tests for burner fuels described elsewhere in this book are also applied to crude oils destined for use as power generation fuels.

Applicable ASTM/IP Standards

ASTM	*IP*	*Title*
D 95	74	Water in Petroleum Products and Bituminous Materials by Distillation
D 96		Water and Sediment in Crude Oils
D 129	61	Sulfur in Petroleum Products by the Bomb Method
D 287		API Gravity of Crude Petroleum and Petroleum Products (Hydrometer Method)
D 323		Vapor Pressure of Petroleum Products (Reid Method)
D 445	71	Kinematic Viscosity of Transparent and Opaque Liquids (and the Calculation of Dynamic Viscosity)
D 473	53	Sediment in Crude and Fuel Oils by Extraction
D 1160		Distillation of Petroleum Products at Reduced Pressure
D 2161		Conversion of Kinematic Viscosity to Saybolt Universal Viscosity or to Saybolt Furol Viscosity
D 2427		Determination of C_2 through C_5 Hydrocarbons in Gasolines by Gas Chromatography
D 2788		Trace Metal in Gas Turbine Fuels (Atomic Absorption Method)
D 2892		Distillation of Crude Petroleum (15 Theoretical Plate Column)
D 4057		Practice for Manual Sampling of Petroleum and Petroleum Products

Bibliography

The Bureau of Mines Routine Method for the Analysis of Petroleum, U.S. Bureau of Mines Bulletin 490.

Moody, G. B., "Origin, Migration, and Occurrence of Petroleum," *Petroleum Exploration Handbook*, McGraw-Hill, New York, Chapter 5.

Nelson, W. L., *Petroleum Refinery Engineering*, 4th ed., McGraw-Hill, New York, 1958.

"Properties of Crude Oils and Liquid Condensate," *Petroleum Production Handbook*, T. C. Frick, Ed., Vol. II, McGraw-Hill, New York, Chapter 18.

2

Gaseous Fuels and Light Hydrocarbons

INTRODUCTION

Natural Gas

THE PRIMARY USE FOR NATURAL GAS is as a fuel for the production of heat. Large quantities are used also as the source of hydrogen in the production of ammonia.

Natural gas is a mixture of hydrocarbon and nonhydrocarbon gases found in porous formations beneath the earth's surface, often in association with crude petroleum. The principal constituent of most natural gases is methane with minor amounts of heavier hydrocarbons and certain nonhydrocarbon gases such as nitrogen, carbon dioxide, hydrogen sulfide, and helium. Raw natural gas may be subclassified as follows:

Nonassociated Gas—Free gas not in contact with significant amounts of crude oil in the reservoir.

Associated Gas—Free gas in contact with crude oil in the reservoir.

Dissolved Gas—Gas in solution with crude oil in the reservoir.

Following processing and conditioning into a merchantable or finished product, natural gas is composed primarily of methane (CH_4) and ethane (C_2H_6).

Natural Gas Liquids (NGL)

Natural gas liquids (NGL) are mixtures of hydrocarbons which are extracted in liquid form from raw natural gas in lease separators, field facilities, or gas processing plants. A significant component of natural gas liquids is condensate. Condensate is similar in characteristics and application to a very light crude oil.

Natural gas liquids may be produced as a "raw mix" or unfractionated stream, fractionated into individual components, or fractionated into various commercial mixtures including liquefied petroleum gas (LP gas) and natural gasoline.

The individual components and commercial products which may be fractionated from natural gas liquids include:

Ethane—A normally gaseous paraffinic compound (C_2H_6) having a boiling point of approximately −88°C (−127°F). Ethane may be handled as a liquid at very high pressures and at temperatures below 32°C (90°F).

Virtually all ethane produced as a separate liquid product are used as a feedstock for the production of ethylene—a basic petrochemical whose applications are covered more fully in the section on olefins and diolefins.

Propane—A normally gaseous paraffinic hydrocarbon (C_3H_8) having a boiling point of approximately −42°C (−44°F). It may be handled as a liquid at ambient temperatures and moderate pressures. Commercial propanes may include varying amounts of ethane, butanes, and liquefied refinery gases.

Butane—A normally gaseous paraffinic hydrocarbon (C_4H_{10}) which is handled usually as a liquid at ambient temperatures and moderate pressures. Butane may be fractionated further into normal butane, which has a boiling point of approximately 0°C (32°F), and isobutane, which has a boiling point of approximately −12°C (11°F).

Liquefied Petroleum Gas (LP Gas)—A mixture of hydrocarbons, principally propane and butanes obtained from either natural gas liquids or refinery gases, which can be stored and handled as liquids at ambient temperatures and moderate pressures. Commercial LP gas also may contain small quantities of ethane and varying quantities of propylene or butylene or both from refinery gases.

The original use for liquefied petroleum gas was as a fuel gas in areas where pipline utility gas was not available. Later, these materials were fractionated into their separate components, principally propane and butane, as preferred feedstocks for the petrochemical industry.

Natural Gasoline—A mixture of hydrocarbons (extracted from natural gas) that consists mostly of pentanes and hydrocarbons which are heavier than pentane.

In the beginning of the natural gasoline industry, the only use for natural gasoline was as motor fuel or as a blending agent in the production of motor fuel. Later, the individual components of natural gasoline, namely isobutanes, butanes, pentanes, and isopentanes, were separated as base stocks for reforming, alkylation, synthetic rubber, and other petrochemicals.

Liquefied Refinery Gas (LR Gas)

Liquefied refinery gases are mixtures of hydrocarbons which are recovered during the refining of crude oil. These materials are composed principally of propane, butanes, propylene, and butylenes, and they can be stored and handled at ambient temperatures and at moderate pressures.

Olefins and Diolefins

Many light olefins and diolefins which are produced in the refinery are isolated for petrochemical use. The individual products are:

Ethylene—A normally gaseous olefinic compound (C_2H_4) having a boiling point of approximately −104°C (−155°F). It may be handled as a liquid at very high pressures and low temperatures. Ethylene is made normally by cracking an ethane or naptha feedstock in a high-temperature furnace and subsequent isolation from other components by distillation.

The major uses of ethylene are in the production of ethylene oxide, ethylene dichloride, and the polyethylenes. Other uses include the coloring of fruit, EP rubbers, ethyl alcohol, and medicine (anesthetic).

Propylene Concentrates—A mixture of propylene and other hydrocarbons, principally propane and trace quantities of ethylene, butylenes, and butanes. Propylene concentrates may vary in propylene content from 70 percent mol up to over 95 percent mol and may be handled as a liquid at normal temperatures and moderate pressures. Propylene concentrates are isolated from the furnace products mentioned in the preceding paragraph on ethylene. Higher purity propylene streams are further purified by distillation and extractive techniques. Propylene concentrates are used in the production of propylene oxide, isopropyl alcohol, polypropylene, and the synthesis of isoprene.

Butylene Concentrates—These are mixtures of butene-1, cis- and trans-butene-2, and, sometimes, isobutene (2-methyl propene) (C_4H_8). These products are stored as liquids at ambient temperatures and moderate pressures. Various impurities such as butane, butadiene, and C_5's generally are found in butylene concentrates.

Virtually all of the butylene concentrates are used as a feedstock for either: (1) an alkylation plant, where isobutane and butylenes are reacted in the presence of either sulfuric acid or hydrofluoric acid to form a mixture of C_7 to C_9 paraffins used in gasoline, or (2) butylene dehydrogenation reactors at butadiene production facilities.

Butadiene—A normally gaseous hydrocarbon (C_4H_6) having a boiling point of −4.38°C (24.1°F). It may be handled as a liquid at moderate pressure. Ambient temperatures are generally used for long-term storage due to the easy formation of butadiene dimer (4-vinyl cyclohexene-1). Butadiene is produced by two major methods: the catalytic dehydrogenation of butane or butylenes or both, and as a by-product from the production of ethylene. In either case, the butadiene must be isolated from other components by extractive distillation techniques and subsequent purification to polymerization-grade specifications by fractional distillation.

The largest end use of butadiene is as

a monomer for production of GR-S synthetic rubber. Butadiene is also chlorinated to form 2-chloro butadiene (chloroprene) used to produce the polychloroprene rubber known as neoprene.

QUALITY CRITERIA

Quality of a product may be defined as its fitness for a purpose. Once the required quality is determined, it is controlled by appropriate testing and analysis. This section will outline quality criteria for significant gaseous fuels and light hydrocarbons. Appropriate American Society for Testing and Materials and Institute of Petroleum (ASTM/IP) methods for testing and analysis will be listed in parenthesis. The full titles of the methods can be found at the end of the chapter under the heading, "Applicable ASTM/IP Standards."

Natural Gas

The principal quality criterion of natural gas is its heating value (ASTM D 1826). In addition, the gas must be readily transportable through high-pressure pipelines. Therefore, the water content, as defined by the water dew point (ASTM D 1142), must be considered to prevent the formation of ice or hydrates in the pipeline. Likewise, the amount of entrained hydrocarbons heavier than ethane, as defined by the hydrocarbon dew point, should be considered to prevent accumulation of condensible liquids that may block the pipeline.

Natural gas and its products of combustion must not be unduly corrosive to the materials with which they come in contact. Thus, the detection and measurement of hydrogen sulfide (ASTM D 2385 or ASTM D 2725) and total sulfur (ASTM D 1072 or ASTM D 3031) are important. The odor of gas must not be objectionable; so, in some cases, mercaptan content (ASTM D 2385) is significant.

If the gas is to be liquefied by cryogenic processing and stored in liquid form, carbon dioxide will separate out of the cold liquid as a solid and interfere with the refrigeration system. Carbon dioxide can be determined by ASTM D 1137 or ASTM D 1945. If the gas is to be used as a feedstock for the production of hydrogen or petrochemicals, a complete analysis (ASTM D 1945) must be made, because some gases contain materials that are deleterious to these processes.

Ethane

Quality criteria for ethane as a petrochemical feedstock are specific to the particular process used for petrochemical manufacture. In general, however, such feedstocks must be very low in moisture content (ASTM D 1142), oxygen (ASTM D 1945), carbon dioxide (ASTM D 1945), sulfur compounds (ASTM D 1072), and other elements that will interfere with the process or the catalysts used in petrochemical manufacture.

Liquefied Petroleum (LP) Gases

For most purposes, liquefied petroleum gas will be stored, handled, and transported as a liquid and used as a gas. In order to handle, store, and transport it safely, the vapor pressure must be known (ASTM D 1267). To be sure that LP gas will convert from liquid to vapor under the intended conditions of use, it is necessary to know something about its volatility (ASTM D 1837). Vapor pressure is an indicator of volatility, but it gives little or no indication of the least volatile components. For this purpose, the results of a weathering test (ASTM D 2158) are needed. The combination of vapor pressure and the results of a weathering test is a good measure of volatility and, therefore, a good indicator of the suitability of LP gas for its intended use.

Copper tubing is used extensively in LP gas systems. As a result, a copper strip corrosion test (ASTM D 1838) is important. Propane vaporizes at temperatures far below the freezing point of water. Therefore, it is important to know that the moisture content of propane (ASTM D 2713) is low enough so that it will not cause freeze-up problems.

Since regulations covering proper filling of LP gas containers are based upon the specific gravity of the liquid, the exact specific gravity of each batch of LP gas must be known (ASTM D 1657).

As with other fuel gases, the products

of combustion should not be unnecessarily corrosive. The common source of corrosive combustion products from LP gas is the sulfur compounds contained in it. Total sulfur content (ASTM D 2784) is a measure of the corrosivity of the combustion product.

When LP gas is vaporized, it is undesirable to leave behind nonvolatile materials that can plug up equipment or cause controls to operate improperly. The residue test (ASTM D 2158) is an indicator of the nonvolatile materials present in LP gas.

When LP gas is intended for use as a petrochemical feedstock, a component analysis is essential. The best and quickest way to analyze such a material is by means of a gas chromatograph (ASTM D 2163/IP 264). This instrument is such a quick, reliable, and accurate tool that it is being used more and more to replace many of the older physical tests. When the composition of an LP gas is known, it is possible to calculate many of its other properties (ASTM D 2598).

Octane number is an important characteristic of LP gas used as engine fuel (ASTM D 1835). Octane number can be determined by the test for knock characteristics (ASTM D 2623/IP 238), or it can be calculated after a chromatographic analysis has been made (ASTM D 2598).

Natural Gasoline

For use as motor fuel or as a component of motor fuel, the primary criteria for the quality of natural gasoline are its volatility and knock performance. The basic measures of volatility are vapor pressure (ASTM D 323) and distillation (ASTM D 216/IP 191). Knock performance is measured by rating in knock test engines by both the motor (ASTM D 2700/IP 236) and research (ASTM D 2699/IP 237) methods.

Other considerations for natural gasoline used in motor fuels are copper corrosion (ASTM D 130/IP 154) and gravity (ASTM D 1298/IP 160). Although gravity has a minor direct relationship with quality, it is necessary to determine gravity for measurement and shipping.

When natural gasoline is used as a feedstock for further processing or petrochemicals, the list of quality criteria is almost endless. For nearly all petrochemical uses composition by hydrocarbon types is needed, and, frequently, a complete analysis of specific components is made (ASTM D 2427). If a catalytic process is involved, total sulfur (ASTM D 1266/IP 191) is very important, because sulfur tends to destroy catalyst activity.

Olefins and Diolefins

Ethylene—Since ethylene is a high-purity product (normally supplied at 99.5 percent mol or higher), the quality criteria of interest are trace components. Those components of greatest concentrations are hydrogen, carbon monoxide, carbon dioxide, oxygen, and acetylene. Generally, moisture and sulfur contents also are quite critical to ethylene-based processes. All of the above impurities are generally catalyst poisons to polymerization processes even in the low mols per million concentration ranges. Therefore, the analytical problems are magnified greatly. To date, ASTM does not have a standard method for moisture at these low levels, but a new trace sulfur method, based on microcoulometry, has been adopted (ASTM D 3246). Ethylene purity is determined by quantitative analysis of the trace impurities and subtracting the sum of their concentrations from 100 percent (ASTM D 2504 and ASTM D 2505).

Propylene Concentrates—Propylene concentrate streams generally require a component analysis, depending upon their final use. The best method is by gas chromatography (ASTM D 2427). Another gas chromatography method is used to identify major impurities (ASTM D 2163/IP 264). Sulfur may be determined by the oxidative microcoulometric technique (ASTM D 3246) just mentioned or, more commonly, by the Wickbold combustion method (ASTM D 2784). As is the case for ethylene, moisture in propylene is critical. Several field tests and a few laboratory tests are in use by individual firms, but no standard method for moisture has been adopted to date. The problems in sampling for moisture content, especially in the less than 10 ppm range, are hard to overcome.

The trace impurities in 90 percent or better propylene, which is used in polymerization processes, become quite critical.

Hydrogen, oxygen, and carbon monoxide are determined by one technique (ASTM D 2504), and acetylene, ethylene, butenes, butadiene, methyl acetylene, and propadiene are determined by using a very sensitive analytical method (ASTM D 2712).

Butylene Concentrates—The major quality criterion for butylene concentrates is the distribution of butylenes which is measured, along with other components, by gas chromatography. Trace impurities generally checked are sulfur (ASTM D 2784), chlorides (ASTM D 2384), and acetylenes. These impurities are sometimes catalyst poisons or become unwanted impurities or both in the final product. When butylene concentrates are used as an alkylation unit feedstock, the diolefin content becomes important.

Butadiene—The major quality criteria for butadiene are the various impurities that may affect the polymerization reactions for which butadiene is used. The gas chromatographic examination (ASTM D 2593/IP 194) of butadiene determines the gross purity as well as C_3, C_4, and C_5 impurities. Most of these hydrocarbons are innocuous to polymerizations, but, some, such as butadiene −1,2 and pentadiene −1,4, are capable of polymer cross-linking.

Acetylenes are determined by using a chemical test method, while carbonyls are determined by the classical hydroxylamine hydrochloride reaction with carbonyls (ASTM D 1089). Total residue is determined to check for nonvolatile matter (ASTM D 1025).

ASTM D 1550, ASTM Butadiene Measurement Tables, provides the data necessary for proper calculation of commercial quantities of butadiene.

Applicable ASTM Specification

D 1835 Specifications for Liquefied Petroleum (LP) Gases

Applicable ASTM/IP Standards

ASTM	*IP*	*Title*
		NATURAL GAS AND ETHANE
D 1070		Relative Density (Specific Gravity) of Gaseous Fuels
D 1072		Total Sulfur in Fuel Gases
D 1142		Water Vapor Content of Gaseous Fuels by Measurement of Dew-Point Temperature
D 1826		Calorific Value of Gases in Natural Gas Range by Continuous Recording Colorimeter
D 1945		Analysis of Natural Gas by Gas Chromatography
D 2385		Hydrogen Sulfide and Mercaptan Sulfur in Natural Gas (Cadmium Sulfate Iodometric Titration Method)
D 2725		Hydrogen Sulfide in Natural Gas (Methylene Blue Method)
D 3031		Total Sulfur in Natural Gas by Hydrogenation
		LIQUEFIED PETROLEUM GAS
D 1265		Sampling Liquefied Petroleum (LP) Gases
D 1267		Vapor Pressure of Liquefied Petroleum (LP) Gases (LP-Gas Method)
D 1657		Density or Relative Density of Light Hydrocarbons by Pressure Hydrometer
D 1837		Volatility of Liquefied Petroleum (LP) Gases
D 1838		Copper Strip Corrosion by Liquefied Petroleum (LP) Gases
D 2158		Residues in Liquefied Petroleum (LP) Gases

Applicable ASTM/IP Standards *(continued)*

ASTM	IP	Title
D 2163	264	Analysis of Liquefied Petroleum (LP) Gases and Propane Concentrates by Gas Chromatography
D 2598		Calculation of Certain Physical Properties of Liquefied Petroleum (LP) Gases from Compositional Analysis
D 2623	238	Knock Characteristics of Liquefied Petroleum (LP) Gases by the Motor (LP) Method
D 2713		Dryness of Propane (Valve Freeze Method)
D 2784		Sulfur in Liquefied Petroleum Gases
		NATURAL GASOLINE
D 130	154	Detection of Copper Corrosion from Petroleum Products by the Copper Strip Tarnish Test
D 216	191	Distillation of Natural Gasoline
D 323		Vapor Pressure of Petroleum Products (Reid Method)
D 1266	101	Sulfur in Petroleum Products (Lamp Method)
D 1298	160	Density, Relative Density (Specific Gravity), or API Gravity of Crude Petroleum and Liquid Petroleum Products by Hydrometer Method
D 2427		Determination of C_2 through C_5 Hydrocarbons in Gasolines by Gas Chromatography
D 2699	237	Knock Characteristics of Motor Fuels by the Research Method
D 2700	236	Knock Characteristics of Motor and Aviation Type Fuels by the Motor Method
		ETHYLENE
D 2504		Noncondensible Gases in C_3 and Lighter Hydrocarbon Products by Gas Chromatography
D 2505		Ethylene, Other Hydrocarbons, and Carbon Dioxide in High-Purity Ethylene by Gas Chromatography
D 3246		Sulfur in Petroleum Gases by Oxidative Microcoulometry
		PROPYLENE CONCENTRATES
D 1265		Sampling Liquefied Petroleum (LP) Gases
D 2163	264	Analysis of Liquefied Petroleum (LP) Gases and Propane Concentrates by GC
D 2504		Noncondensible Gases in C_3 and Lighter Hydrocarbon Products by Gas Chromatography
D 2712		Hydrocarbon Traces in Propylene Concentrates by Gas Chromatography
		BUTYLENE CONCENTRATES
D 1265		Sampling Liquefied Petroleum (LP) Gases
D 2384		Traces of Volatile Chlorides in Butane-Butene Mixtures
D 2784		Sulfur in Liquefied Petroleum Gases (Oxy-Hydrogen Burner or Lamp)

Applicable ASTM/IP Standards *(continued)*

ASTM	*IP*	*Title*
		BUTADIENE
D 1025		Nonvolatile Residue of Polymerization Grade Butadiene
D 1089		Carbonyl Content of Butadiene
D 2426		Butadiene Dimer and Styrene in Butadiene Concentrates by Gas Chromatography
D 2593	194	Butadiene Purity and Hydrocarbon Impurities by Gas Chromatography

Bibliography

Katz, D. L., *Handbook of Natural Gas Engineering*, McGraw-Hill, New York, 1969.

3

Petroleum Solvents

INTRODUCTION

THE TERM "PETROLEUM SOLVENTS" describes the special liquid hydrocarbon fractions obtained from petroleum for use in industrial processes and commercial formulations to dissolve, suspend, or transport the other ingredients of the process or formulation. In recent years, the variety of petroleum solvents has been increased considerably due to the development of refinery processes designed primarily for the transformation of low-octane feedstocks to high-octane fuels. These new developments led to the production of certain important solvents and intermediates which were obtained previously only from the coal carbonizing industry. By definition, the solvents obtained from the petrochemical industry—the alcohols, ethers, etc.—are outside the scope of this chapter.

Petroleum solvents are available for a variety of applications. Modern refining techniques produce solvents with a high degree of purity and stability in a wide range of grades and boiling points. Petroleum solvents are particularly attractive, because they are readily available in large volumes at low costs compared to solvents from other sources.

GENERAL USES

Solvents are used extensively by industry in manufacturing processes for such diverse products as paint, printing ink, polish, adhesives, perfumes, glues, fats, etc. Further uses are found in the dry cleaning, leather, and fur industries and the pesticide field. Solvents, in highly purified condition, also are becoming increasingly important for use as reaction media in certain catalytic processes.

PRODUCT REQUIREMENTS

The variety of applications emphasizes the versatility of petroleum solvents. The main characteristics which determine the suitability of a petroleum fraction for a particular use are its solvent properties, volatility, purity, gravity, odor, toxicity, and air pollution control/limitations.

Solvency

In many applications, petroleum solvents are used as a vehicle to dissolve resins, oils, gums, or waxes. In other cases, the solvent merely suspends such items as pigments, fillers, and water. Examples of this are their use in the printing ink, rubber coating and dipping, paint, lacquer, and polish industries.

Solvent properties are determined by the hydrocarbon types present. In general, aromatic hydrocarbons have the highest solvent power and straight-chain aliphatics the lowest.

A primary characteristic of solvency is the ability of the solvent to dissolve a resin or other film formers. In general, the more polar resins require the more polar solvents, that is, alcohols, esters, and ketones.

The amount of film former that the solvent will dissolve determines the solids content (or nonvolatile content) of the system and ultimately the viscosity and flow characteristics of the finished product. This is the most common basis for rating solvents as to solvency for a particular resin.

As a solvent is added to a resin in increasing amounts, there is a point at which the resin precipitates or "kicks out." This "dilution limit" is expressed as the percent nonvolatile when precipitation occurs. The dilution limit is used frequently as an indication of the relative solvency between solvents.

Volatility

Next to solvency, volatility may be the most important property of a solvent since volatility largely governs evaporation rate or drying time. Volatility is determined by the distillation temperature or range of temperatures for the solvent. Pure hydrocarbons such as pentane, hexane, heptane, benzene, toluene, and xylene are characterized by a narrow boiling range of −17 to −16°C (2 to 3°F). Other petroleum solvents that are mixtures of many hydrocarbons have wide boiling range widths that vary with their composition.

An excellent example of the importance of evaporation rate can be found in the factors considered in the selection of a solvent for coatings. In coatings, evaporation rate influences levelling, flowing, sagging, wet-edge time, and gloss. The optimum evaporation rate varies with the method of application, from fastest for spraying to intermediate for brushing to slowest for flow coating and conveyor.

The choice of a solvent for a given application is dictated also by temperature of use. A high-boiling sample will be required for a heat-set ink where the operating temperature may be as high as 316°C (600°F), but a low-boiling solvent would be selected for fast drying in a foil printing press. Fortunately, solvents are available to satisfy the wide range of requirements for evaporation rates and temperatures.

Purity

The purity of the various hydrocarbon solvents pertains not only to the minor concentrations of nonhydrocarbon contaminants but, in some cases, particularly with the highly purified lower boiling aromatics, to the presence of other undesirable hydrocarbons.

Petroleum solvents, other than hydrocarbons such as toluene, are mixtures of straight, branched-chain, and cyclic paraffins and aromatic hydrocarbons with a possible trace of olefinic material. With this composition, they are, in the refined state, inert substances. As inert materials, solvents may be used in the preparation of surface coatings and adhesives and in a multitude of applications without risk of side effects due to reaction with other substances in the formulation or with the application surface.

The importance of any particular impurity will be a function of the solvent's application. As a result, purchasers generally set solvent specifications. Normally, part of these specifications will contain limits on the maximum level of specific contaminants germane to their processes.

Gravity

Gravity, as defined later, is proportional to density. Since costs calculations are based on density, this solvent property becomes very important in the marketplace.

Odor

The odor of a solvent generally refers to the odor of the vapor during and shortly following application. Occasionally, there may be a persistent residual odor contributed by trace contaminants.

Ordinarily, the odor of industrial product finishes is not critical because finishers become accustomed to strong smelling solvents, and most of the vapors are removed by exhaust systems. However, finishers are likely to object if the paint from one supplier has a different odor than that obtained from another.

Odor is most critical for interior trade sales products such as paints used in homes. Even the mild odor of the typical aliphatic hydrocarbon solvents frequently is considered unpleasant. This has contributed to the widespread acceptance of latex paints which, by comparison, have subdued odors.

In the case of paints, chemists have little leeway regarding odor, since the solvents must be chosen primarily for solvency and evaporation rate. To date, the numerous "deodorants" promoted for use in paint have been of limited value. A partial solution is the use of mineral spirits which are almost odor free.

Toxicity

The toxicological properties of the various hydrocarbon solvents are quite important. The threshold limit values (TLVs) adopted by the American Conference of Governmental Industrial Hygienists (ACGIH) limit the use of solvents in high concentration in workroom air. In some cases, the choice between different solvent systems is based solely on TLVs. (Threshold limit values for individual solvents are available from ACGIH.)

Toxicity of solvents pertains to the potential injury to health that could occur if the solvent gets into the bloodstream through inhalation. As an example, benzene is sufficiently toxic to prohibit its use in federal paint specifications. For this reason, paint manufacturers discourage the use of benzene in paint and paint remover formulations.

As another example, toluene and xylene are more toxic than the aliphatic hydrocarbons in the same boiling range. Therefore, aliphatic systems are preferred to toluene and xylene if the loss of solvency can be made up by the use of oxygen containing compounds.

Solvents used in paints for home use are relatively nontoxic. However, prolonged breathing of concentrated solvent vapors should be avoided.

Air Pollution Requirements

The ability of a solvent to meet air pollution requirements has become an increasingly important consideration. Most air pollution regulations prescribe the quantity of a given solvent that may be discharged into the air over a given time interval and the composition of allowable solvent blends.

Los Angeles Rule 66, which emphasizes the use of non- or low-aromatic hydrocarbons, has served as a guide for other regions and municipalities. Limitations under Rule 66 are not based on toxicity but on the susceptibility of the solvents to undergo photochemical reactions which cause smog and eye irritation. Rule 66 limits the use of toluene to 20 percent of the solvent and xylenes to 8 percent (by volume).

TYPES OF HYDROCARBONS

The hydrocarbons may be divided into subgroups in different ways based on chemical composition. The division that promotes the easiest understanding is:

1. Aliphatics.
   (*a*) Normal paraffins, C_nH_{2n+2}.
   (*b*) Isoparaffins, C_nH_{2n+2}.
2. Naphthenes (cycloparaffins), C_nH_{2n}.
3. Aromatics, C_nH_{2n-6}.
4. Olefins, C_nH_{2n}.

Each of these divisions will be discussed in subsequent paragraphs. Corresponding structural formulas, based on compounds having six carbon atoms, are shown in Fig. 1.

Aliphatics

There are two types of open chain or aliphatic hydrocarbon solvents: normal paraffins and isoparaffins. The normal paraffins have a straight chain of carbon atoms, while the isoparaffins have side chains. Commercial solvents are in the range of compounds with 6 to 15 carbons with the most widely used solvents being compounds with 7 to 12 carbons.

Both types of paraffins conform to the principle that increasing molecular weight results in lower solvency and lower volatility. The normal paraffins are characterized by low solvency and mild odor. As implied by the name, the isoparaffins are isomers of the normal paraffins which they resemble in solvency. Their only distinctive and significant property is extremely low odor. They are almost odorless in the C_9 and higher range.

Naphthenes (Cycloparaffins)

The only naphthenes that are present in paint solvents in significant quantity are cyclopentane, cyclohexane, and their alkyl substituted homologues. The amount of naphthenes in nonaromatic hydrocarbon solvents varies greatly—from 0 to 90 percent by volume. Because of their cyclic structure, the naphthenes are less volatile than paraffins with the same number of carbons and have stronger solvency. They

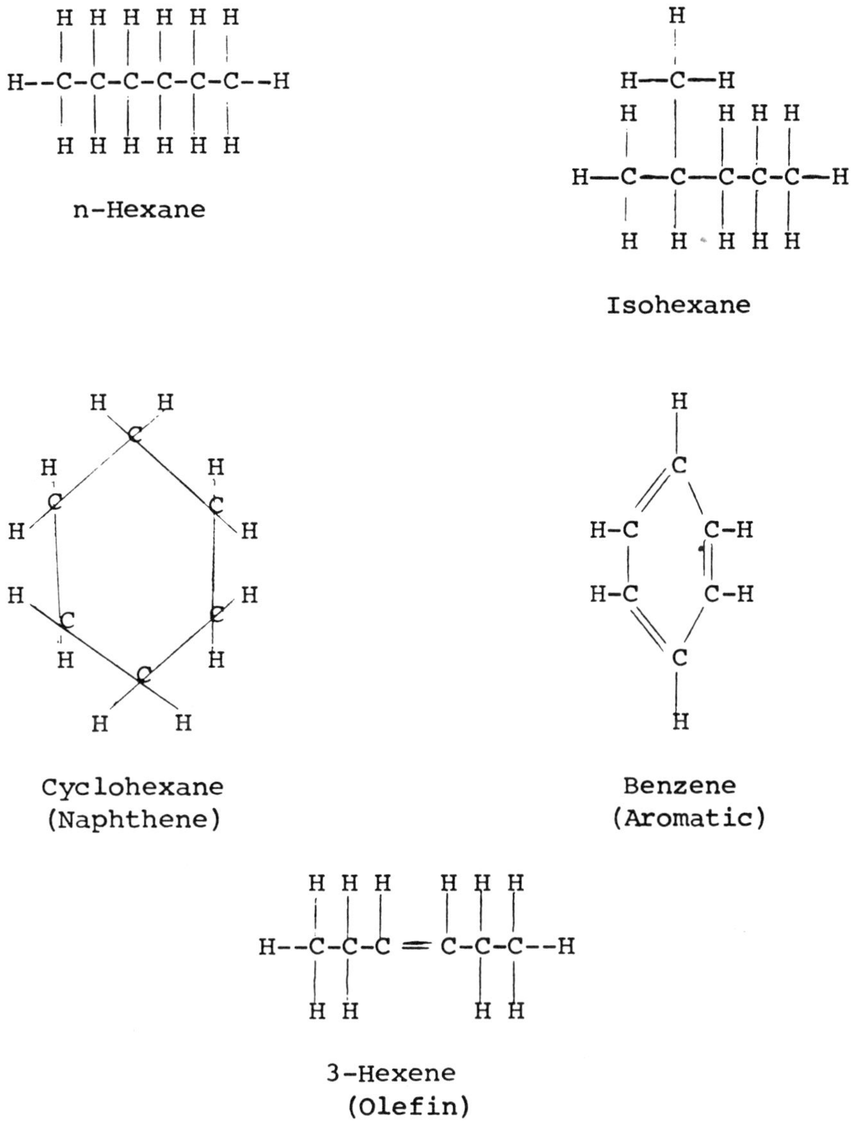

n-Hexane

Isohexane

Cyclohexane
(Naphthene)

Benzene
(Aromatic)

3-Hexene
(Olefin)

FIG. 1—Chemical structures of some C_6 hydrocarbons.

possess an odor similar to that of the terpenes which eliminates them from low-odor solvents.

Aromatics

All of the aromatics are based on the unsaturated, six carbon ring structure of benezene as shown in Fig. 2. The compounds of higher molecular weight are called alkyl benzenes because they are benzene with one or more hydrogens replaced by alkyl groups, such as methyl, ethyl, or butyl. This is shown in Fig. 2 for toluene and ortho-xylene, one of the three xylene isomers.

The more important products that are higher than xylene in molecular weight are n-propylbenzene, iso-propylbenzene (cumene), n-butylbenzene, and iso-butylbenzene. Mixtures of these higher molecular weight compounds are sold as a solvent under the name Hi-flash Solvent Naphtha.

Outstanding properties of the aromatic solvents, in comparison with aliphatic hydrocarbons, are much higher solvency and stronger odor. Although naphthenes are a minor factor, it is not too strong to say that the solvent power of commercial hydrocarbon solvents is roughly in proportion to their aromatic content.

The almost pure aromatic hydrocarbons are correctly designated benzene, toluene, and xylene. Commercial grades with distillation ranges up to 10°C are equally satisfactory for use in coatings. These ma-

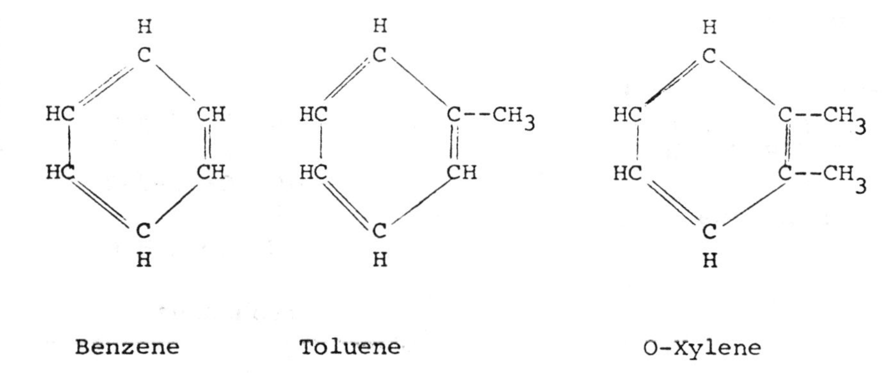

FIG. 2—Structure of lower aromatic hydrocarbons.

terials are known as benzol, toluol, and xylol.

COMMERCIAL HYDROCARBON SOLVENTS

General Solvent Types

The foregoing discussion of hydrocarbon solvents by chemical types provides a background for the commercial products and their usage. While a paint chemist should have some interest in the chemical constituents of solvents, his primary concern must be with the commercial solvents as purchased. Of necessity, his selection and use of solvents is based solely on the properties of commercial products as determined by tests. Solvent users are guided mainly by the following commercial classification of hydrocarbon solvents:

1. Narrow cut aliphatics.
2. Special boiling range solvents.
3. Mineral spirit types.
4. Aromatics.
5. Kerosine.

The approximate properties of these typical hydrocarbon solvents are shown in Table 1.

Specific Types of Solvents

Although industrial users of solvents are guided by the commercial classification just outlined, there are specific types of commercial solvents that merit a thumbnail description.

Hexane and Heptane—Hexane and heptane are used frequently in multistage processes where the solvent must be still present in one step, but be gone completely before a subsequent step. They are also used as solvents and suspending agents in the petrochemical industry where they are recovered and reused. Heptane is used in place of hexane when a higher flash point or higher boiling range solvent is needed.

Petroleum Ether—Petroleum ether is the fastest evaporating of the commonly used solvents and therefore has obvious advantages, but its flash point is correspondingly low. It is used as a solvent for oils, fats, waxes, in paints and varnishes, and as a fuel.

"Textile" Spirits—This close cut solvent is predominantly hexane. It is used only when very fast setting is desired, but since it has a broader boiling range than hexane, evaporation occurs over a longer time interval.

Lacquer Diluent—Lacquer diluent is designed primarily for use in nitrocellulose lacquer as a replacement, or partial replacement, for toluene. Lacquer diluent is used in place of toluene because it is cheaper. The quality of lacquer diluent is proportional directly to aromatic content. The more aromatics, the better the quality. Lacquer diluents are used also as solvents in gravure inks for printing.

VM&P Naphtha—Large quantities of VM&P naphtha are used. It resembles regular mineral spirits in hydrocarbon type composition and has about the same solvency. However, it is lower boiling, has a lower flash point, and evaporates much faster. The fast evaporation and low-flash point make it generally unsuitable for interior architectural finishes, but it is well adapted for application by spraying and, in some cases, by dipping. The grades with an initial boiling point of about 112°C (235°F)

TABLE 1. Approximate properties of typical hydrocarbon solvents.

	Distillation Range, °C (°F)		Evaporation Rate N=BuAc=100	Flash Point, Tag Closed Cup, °C (°F)	Kauri-Butanol Value	Aniline Point, °C (°F)	Aromatics Volume %	Sp Gr 60/60
	Initial	Dry Point						
				NARROW CUT ALIPHATICS				
Hexane	66 (151)	69 (157)	>1000	<−15 (<0)	31	63 (146)	<1	0.680
Heptane	94 (201)	98 (209)	600	−5 (23)	38	53 (127)	3	0.728
				SPECIAL BOILING POINT SOLVENTS				
Petroleum ether	35 (95)	60 (140)	>1000	15 (20)	26	66 (150)	0	0.645
Textile spirits	63 (145)	79 (175)	850	−5 (20)	34	57 (134)	5	0.690
Rubber solvent	62 (144)	120 (248)	700	<−18 (<0)	36	55 (131)	5	0.712
Lacquer diluent	93 (200)	104 (220)	400	−7 (20)	40	49 (120)	15	0.728
VM&P naphtha	99 (210)	149 (300)	150	7 (45)	38	54 (130)	10	0.752
Ink oil	257 (495)	293 (560)	<1	116 (240)	26	80 (176)	<5	0.820
				MINERAL SPIRITS				
Regular (Stoddard)	160 (320)	193 (380)	10	42 (108)	39	54 (130)	15	0.789
140-flash point	182 (360)	204 (400)	5	60 (140)	35	62 (143)	15	0.792
Odorless	177 (350)	193 (380)	7	52 (125)	27	84 (184)	<1	0.759
High solvency	160 (320)	193 (380)	8	43 (110)	44	41 (105)	30	0.808
				AROMATICS				
Benzene	79 (175)	81 (177)	500	<−12 (<10)	115	14 (57)[a]	99+	0.885
Toluene	110 (230)	111 (232)	200	4 (40)	105	8 (46)[a]	99+	0.870
Xylene	139 (282)	141 (285)	60	28 (83)	98	11 (52)[a]	99+	0.869
High-flash aromatic naphtha	154 (310)	177 (350)	19	42 (107)	92	13 (56)[a]	98+	0.874
High-flash aromatic naphtha	182 (360)	210 (410)	4	66 (151)	90	18 (64)[a]	97+	0.891
Heavy aromatic naphtha	182 (360)	282 (540)	1	67 (152)	105	24 (76)[a]	86+	0.933
				KEROSINE				
Regular	177 (359)	268 (515)	1	60 (140)	30	71 (160)	<25	0.804
Odorless	204 (400)	249 (480)	1	77 (170)	27	88 (190)	<1	0.782

[a]Mixed aniline point, °C (°F).

or higher are differentiated frequently as high-flash VM&P naphtha.

Ink Oils—Ink oils are used in place of kerosines where narrow-cut ink solvents are desired. The narrow cut provides better drying control of the ink.

Mineral Spirits—Mineral spirits constitute about three fourths of all hydrocarbon solvents used in the paint industry and 90 percent of the solvents in architectural finishes. It comprises four general types: (1) regular, (2) odorless, (3) high flash, and (4) high solvency. Within each of these types, there are products that offer a choice in boiling range and volatility, and thus permit some control of "setting" rate, brushability, and wet-edge time.

Benzene (Benzol)—The use of benzene in coatings is considered bad practice because the vapors are toxic. Its only common use by the paint industry is in solvent type paint and varnish removers where it is more effective than the higher-boiling aromatics.

Toluene (Toluol)—Toluene is used widely when maximum solvency and rather fast evaporation are desired. It is the best fast diluent for nitrocellulose lacquers and also is used in synthetic enamels of several types. The industrial pure grades with distillation ranges of a few degrees centigrade give the same results as the 1°C nitration grade.

Xylene (Xylol)—Xylene is essentially a mixture of the ortho, meta, and para isomers of xylene with ethylbenzene. It is used instead of toluene where slower evaporation is sought. In both lacquers and synthetic enamels, xylene is used frequently in admixture with toluene. It is the sole or main hydrocarbon solvent in many synthetic industrial enamels that are applied by spraying. Unlike toluene, xylene does not require a red label for interstate shipments.

Hi-Flash Solvent Naphtha—Formerly this term referred only to a high-boiling fraction obtained from coal tar. Today, it also covers an equivalent product that is derived from petroleum. Many industrial paints require at least a portion of a strong hydrocarbon solvent that evaporates slower than xylene. Hi-flash Solvent Naphtha fills this need.

Kerosine—Kerosine is the only hydrocarbon solvent that is rated as having a wide distillation range. It is used when extremely slow setting and rather low solvency are desired, and the kerosine odor is permissible. Two such purposes are wood fillers and putties. In most applications, kerosine is combined with a larger amount of mineral spirits.

TEST METHODS FOR SOLVENTS

The most important tests conducted on solvents are shown in Table 2. These tests are of two general types. Control tests are those used to test shipments for uniformity and compliance with specifications, and evaluation tests are those conducted on new materials to determine their properties and suitability for use. As a rule, the control tests are more simple. Some tests may be used for both purposes.

The following paragraphs address sampling techniques for petroleum solvents and discuss the purposes and principles of the tests listed in Table 2. Specific details on the various tests can be found in the *Annual Book of ASTM Standards*, Volumes 05.01, 05.02, 06.02, 06.03, 10.03, and 15.05.

Sampling

Because of the high standards set for petroleum solvents, it is essential to employ the correct techniques when taking samples for tests. Mishandling, or the slightest trace of contaminant, can give rise to misleading results. Sampling methods are described in ASTM Practice for Manual Sampling of Petroleum and Petroleum Products (D 4057). Special care is necessary to ensure that containers are scrupulously clean and free from odor. Samples should be taken with the minimum of disturbance so as to avoid loss of volatile components. In the case of the lightest solvents, it may be necessary to chill the sample.

While awaiting examination, samples should be kept in a cool, dark place to ensure that they do not discolor or develop odors.

TABLE 2. ASTM tests commonly conducted on solvents.

Specific gravity	D 891 A or B
Distillation	
(*a*) petroleum type solvents	D 86
(*b*) Other solvents	D 1078
Flash point	
(*a*) tag closed tester	D 56
(*b*) tag open tester	D 1310
Evaporation rate	pending
Kauri-butanol value of hydrocarbon solvents	D 1133
Copper corrosion, aromatic hydrocarbons	D 849
Dilution ratio in nitrocellulose solutions	D 1720
Viscosity of nitrocellulose solutions	D 1343

Specific Gravity

The determination of specific gravity serves two purposes: it provides a check on the uniformity of shipments, and it permits calculation of the weight per gallon.

Specific gravity is defined as the ratio of a weight of a given volume of material to the weight of an equal volume of water under specified conditions. Specific gravity for petroleum solvents is usually determined by ASTM Test Methods for Specific Gravity of Liquid Industrial Chemicals (D 891). A gravity balance or a glass hydrometer is used with this procedure.

Producers of petroleum solvents frequently use American Petroleum Institute (API) gravity instead of specific gravity. API gravity was developed to permit the use of whole numbers.

To avoid dealing with two kinds of gravity, paint laboratories usually convert the API values to specific gravity by use of conversion tables or by the formula

API gravity, deg

$$= \frac{141.5}{\text{sp gr at } 15.5/15°\text{C } (60/60°\text{F})} \; 131.5$$

For figuring formula yields and important composition relationships, paint laboratories must have a list of raw materials that includes the weight per gallon of all liquid materials at normal operating temperature. When the specific gravity of solvents is determined at the customary 15.5/15.5°C (60/60°F), a temperature conversion factor must be used to obtain the specific gravity at 21°C (70°F), and this factor differs with the solvent. This corrected specific gravity must then be converted to pounds per gallon.

Distillation

The significance of distillation results is based on the close relationship to volatility which, in turn, largely governs evaporation rate. ASTM Test for Distillation of Petroleum Products (D 86/IP 123) is used for solvents with a wide distillation range (this includes the common aliphatic and aromatic naphtha solvents). ASTM Test for Distillation Range of Volatile Organic Liquids (D 1078) is the procedure used for solvents with a narrow range of distillation (aromatic hydrocarbons, alcohols, esters, and ketones). Regardless of solvent type, a very narrow distillation range has no direct advantage in coatings. However, it may have significance as an indication of the degree of purity of oxygenated solvents.

Flash Point

The flash point of a liquid is the lowest temperature at which application of a flame causes the vapor above the sample to ignite. Flash point is indicative of potential fire hazard. The most common method of determining flash point is with the ASTM Test Method for Flash Point by Tag Closed Tester (D 56) which confines the vapor until the instant the flame is applied. However, Interstate Commerce Commission regulations on the shipping of flammable solvents are based on the ASTM Flash Point of Liquids by Tag Open-Cup Apparatus (D 1310) which gives slightly higher values.

In mixtures of solvents and in coatings, the flash point is substantially that of the lowest flash portion that is present in an appreciable amount.

Evaporation Rate

Although there is a significant relation between distillation range and evaporation rate, the relationship is not direct. Because of its importance, several methods of measuring evaporation rates of straight solvents (not the solvent in a film coating) have been developed by various commercial and governmental organizations.

A simple procedure for evaporation rate that can be conducted in any laboratory with sufficient precision for practical purposes is as follows:

> Use tared 1-ounce shallow salve cans with covers and do weighing on an analytical balance. Weigh 1 gram of test solvent in one can and 1 gram of n-butyl acetate in another can. Place the containers in a draft-free area, preferably with controlled temperature and humidity. Remove the covers and reweigh at intervals until the samples have completely evaporated. The time in minutes for n-butyl acetate divided by the time for the test solvent is the evaporation rate. It may be advisable to change the size of the samples for solvents that are quite fast or quite slow. A very slow tail of a few percent may be disregarded.

Evaporation rates are usually expressed relative to n-butyl acetate, which is assigned a value of one. A value of 0.4 means that the solvent evaporates 0.4 times faster than butyl acetate and a value of 2.5 means 2.5 times faster. Since the last 5 percent of a solvent may show a definite lag in evaporation but has small influence on performance, it is sometimes disregarded in assigning the rate.

There is a formula for calculating approximate evaporation rates from distillation temperatures. However, solvent producers supply information on the individual solvents, so the only possible need for such calculations is for solvent mixtures. In view of the complexity of the calculations, direct determination probably would take no longer and be no less reliable.

Kauri-Butanol Value

ASTM Test for Kauri-Butanol Value of Hydrocarbon Solvents (D 1133) is used to determine relative solvent power. The kauri-butanol (K-B) value is the number of milliliters of the solvent, at 15°C (77°F), required to produce a defined degree of turbidity when added to 20 g of a standard solution of gum kauri resin in normal butyl alcohol. For K-B values of 60 and over, the standard is toluene which has an assigned value of 105. For K-B values under 60, the standard is a blend of 75 percent n-heptane and 25 percent toluene which has an assigned value of 40. Higher values indicate greater solvent power. The K-B value of products that are classified as regular mineral spirits normally varies between 34 and 44; xylene is 93, and the aromatic naphtha solvents range all the way from 55 to 108.

Aniline Point and Mixed Aniline Point

ASTM Test Method for Aniline Point and Mixed Aniline Point of Petroleum Products and Hydrocarbon Solvents is another method for determining the solvent power of hydrocarbon solvents. It is a more precise technique than the method for K-B value, and the results have better correlation with performance in coatings. The two methods may indicate a reversal of solvent power when applied to solvents that are in the same range. Therefore, in evaluation testing it is best to employ both methods. For control purposes, one method should be adequate, since there is no significant difference in repeatability.

Sulfur

Copper strip corrosion tests are used to detect the presence and determine the relative amounts of sulfur compounds in solvents. Sulfur compounds are objectionable because they impart odor and may cause darkening of cooked varnishes and resins.

ASTM Test for Copper Corrosion of Industrial Aromatic Hydrocarbons (D 849) is used for the determination of sulfur compounds in aromatic hydrocarbons while ASTM Method for Detection of Copper Corrosion from Petroleum Products by the Copper Strip Tarnish Test is applicable for sulfur compounds in aliphatic hydrocarbons. With modern refining methods, copper corrosion tests have assumed less importance. They are now considered optional even for control testing.

Dilution Ratio in Nitrocellulose Solutions

Solvent mixtures for nitrocellulose lacquers contain three classes of volatile liquids: active solvents, latent solvents (alcohols), and diluents. The active and latent solvents may be grouped together under the term "oxygenated solvents." The diluents are hydrocarbon solvents which have no solvency for nitrocellulose and employed mainly to reduce cost without sacrifice of quality. The proportion of hydrocarbon solvents that can be used without precipitation of the nitrocellulose differs with the kind of hydrocarbon. The dilution ratio test determines this proportion for particular hydrocarbon solvents. The dilution ratio is

$$\frac{\text{volume of hydrocarbon solvent}}{\text{volume of butyl acetate}} \quad \text{at the end point}$$

This ratio is roughly proportional to the aromatic content of the hydrocarbon.

The dilution ratio test also is usefully applied to the oxygenated solvents. Oxygenated solvents differ in the amount of diluent that they will tolerate. This affects the cost of formulating with a particular oxygenated solvent. For this purpose, a single diluent, toluene, is employed, and the amount is varied with each oxygenated solvent being tested.

Viscosity of Nitrocellulose Solutions

One of the disadvantages of nitrocellulose in coatings is that the more durable grades produce solutions that have high viscosity and, therefore, low solids at a viscosity suitable for application. The viscosity of a solution is affected by the kind of solvent or mixture of solvents. ASTM Test for Viscosity of Cellulose Derivatives by Ball-Drop Method (D 1343) is designed to determine the inherent viscosity characteristic of various grades of nitrocellulose. The test is equally suitable for comparing the viscosity effect of solvents by varying the solvent while holding the kind and amount of nitrocellulose constant.

Applicable ASTM Specifications

Aliphatic Hydrocarbons

D 235	Mineral Spirits (Petroleum Spirits) (Hydrocarbon Dry Cleaning Spirits)
D 1836	Commercial Hexanes

Aromatic Hydrocarbons

D 362	Toluene, Industrial Grade
D 835	Benzene, 48.5, Refined (Nitration Grade)
D 836	Benzene, Industrial Grade
D 841	Toluene, Nitration Grade
D 843	Xylene, Nitration Grade
D 846	Xylene, Ten Degree
D 2359	Benzene-535, Refined

Applicable ASTM/IP Standards

Physical Properties

ASTM	*IP*	*Title*
		VOLATILITY
Distillation		
D 86	123	Distillation of Petroleum Products
D 850		Distillation of Industrial Aromatic Hydrocarbons and Related Materials
D 1078		Distillation Range of Volatile Organic Liquids
D 1160		Distillation of Petroleum Products at Reduced Pressure
D 2887		Boiling Range Distribution of Petroleum Fractions by GC

Applicable ASTM/IP Standards

Physical Properties (continued)

ASTM	*IP*	*Title*
Flash Point		
D 56		Flash Point by Tag Closed Tester
D 92	36	Flash and Fire Points by Cleveland Open Cup
D 93	34	Flash Point by Pensky-Martin Closed Tester
D 1310		Flash Point and Fire Points of Liquids by Tag Open-Cup Apparatus
Autoignition Temperature		
E 659		Autoignition Temperature of Liquid Chemicals
Residue		
D 1353		Nonvolatile Matter in Volatile Solvents for Use in Paint, Varnish, Lacquer, and Related Products
		DENSITY AND SPECIFIC GRAVITY
D 287		API Gravity of Crude Petroleum and Petroleum Products (Hydrometer Method)
D 891		Specific Gravity of Liquid Industrial Chemicals
D 941		Density and Relative Density (Specific Gravity) of Liquids by Lipkin Bicapillary Pycnometer
D 1217		Density and Relative Density (Specific Gravity) of Liquids by Bingham Pycnometer
D 1298	160	Density, Relative Density (Specific Gravity), or API Gravity of Crude Petroleum and Liquid Petroleum Products by Hydrometer Method
D 1555		Calculation of Volume and Weight of Industrial Aromatic Hydrocarbons
D 2935		Apparent Density of Industrial Aromatic Hydrocarbons
		FREEZING POINT
D 852		Solidification Point of Benzene
D 1015		Freezing Points of High-Purity Hydrocarbons
D 1493		Solidification Point of Industrial Organic Chemicals
D 2500		Cloud Point
		VISCOSITY
D 88		Saybolt Viscosity
D 445	71	Kinematic Viscosity of Transparent and Opaque Liquids (and the Calculation of Dynamic Viscosity)
D 2161		Conversion of Kinematic Viscosity to Saybolt Universal Viscosity or to Saybolt Furol Viscosity
		COLOR
D 156		Saybolt Color of Petroleum Products (Saybolt Chromometer Method)
D 848		Acid Wash Color of Industrial Aromatic Hydrocarbons

Applicable ASTM/IP Standards

Physical Properties (continued)

ASTM	*IP*	*Title*
D 1209		Color of Clear Liquids (Platinum-Cobalt Scale)
		REFRACTIVE INDEX
D 1218		Refractive Index and Refractive Dispersion of Hydrocarbon Liquids
		SOLVENCY
D 611		Aniline Point and Mixed Aniline Point of Petroleum Products and Hydrocarbon Solvents
D 1133		Kauri-Butanol Value of Hydrocarbon Solvents
D 1343		Viscosity of Cellulose Derivatives by Ball-Drop Method
D 1720		Dilution Ratio of Active Solvents in Cellulose Nitrate Solutions
		ODOR
D 1296		Odor of Volatile Solvents and Diluents
		DEFINITIONS AND HANDLING
D 268		Sampling and Testing Volatile Solvents and Chemical Intermediates for Use in Paint and Related Coatings and Materials
D 1086		Measuring Temperature of Petroleum and Petroleum Products
D 1250	200	Petroleum Measurement Tables
D 4057		Practice for Manual Sampling of Petroleum and Petroleum Products

Chemical Properties

ASTM	*IP*	*Title*
		BULK COMPOSITION
D 1016		Purity of Hydrocarbons from Freezing Points
D 1319	156	Hydrocarbon Types in Liquid Petroleum Products by FIA
D 2008		Ultraviolet Absorbance and Absorptivity of Petroleum Products
D 2268		Analysis of High-Purity N-Heptane and *Iso*octane by Capillary GC
D 2306		Xylene Isomer Analysis by GC
		IMPURITY DETERMINATIONS
Hydrocarbon Types		
D 611		Test Method for Aniline Point and Mixed Aniline Point of Petroleum Products and Hydrocarbon Solvents
D 875		Calculation of Olefins and Aromatics in Petroleum Distillates from Bromine Number and Acid Absorption (Intent to Withdraw)

Applicable ASTM/IP Standards

Physical Properties (continued)

ASTM	IP	Title
D 1159	130	Bromine Number of Petroleum Distillates and Commercial Aliphatic Olefins by Electrometric Titration
D 2267		Aromatics in Light Naphthas, and Aviation Gasolines by GC
D 2360		Trace Impurities in Monocyclic Aromatic Hydrocarbons by GC
D 2600	262	Aromatic Traces in Light Saturated Hydrocarbons by GC
D 2710		Bromine Index of Petroleum Hydrocarbons by Electrometric Titration
Sulfur		
D 130	154	Detection of Copper Corrosion from Petroleum Products by Copper Strip Tarnish Test
D 849		Copper Corrosion of Industrial Aromatic Hydrocarbons
D 853		Hydrogen Sulfide and Sulfur Dioxide Content (Qualitative) of Industrial Aromatic Hydrocarbons
D 1266	107	Sulfur in Petroleum Products (Lamp Method)
D 2324		Determination of Carbon Disulfide in Benzene
D 3120		Trace Quantities of Sulfur in Light Liquid Petroleum Hydrocarbons by Oxidative Microcoulometry
Acidity		
D 847		Acidity of Benzene, Toluene, Xylenes, Solvent Naphthas and Similar Industrial Aromatic Hydrocarbons
D 1093		Acidity of Distillation Residues or Hydrocarbon Liquids
D 2896		Total Base Number of Petroleum Products by Potentiometric Perchloric Acid Titration
Water		
E 203		Water Using Karl Fischer Reagent
D 1364		Water in Volatile Solvents (Fischer Reagent Titration Method)
D 1744		Water in Liquid Petroleum Products by Karl Fischer Reagent

Bibliography

Durrans, T. H., *Solvents*, Chapman & Hall, London, 1957.

Marsden, L., *Solvents Guide*, Cleaver-Hume, London, 2nd edition, 1963.

Mellan, I., *Handbook of Solvents*, Vol. 1, Reinhold, New York, 1957.

Mellan, I., *Industrial Solvents*, Reinhold, New York, 2nd edition, 1950.

Mellan, I., *Industrial Solvents Handbook*, Noyes Data Corporation, Newark, New Jersey, 1970.

Modern Petroleum Technology, Institute of Petroleum, London, 3rd edition, 1962.

Petroleum Products Handbook, V. B. Guthrie, Ed., McGraw-Hill, New York, 1960.

Reynolds, W. W., *Physical Chemistry of Petroleum Solvents*, Reinhold, New York, 1963.

Automotive Gasolines

4

INTRODUCTION

Automotive gasolines are used to fuel internal combustion spark-ignition engines. While gasolines discussed in this chapter are used primarily in passenger car and highway truck service, they also are used in off-highway utility vehicles, farm machinery, two- and four-stroke cycle marine engines, and in other spark-ignition engines employed in a variety of different service applications.

Automotive gasolines essentially are blends of hydrocarbons derived from petroleum. In addition, they may contain selected additives that impart specific features to the finished gasoline. The hydrocarbons are derived from fractional distillation of crude oil and from complex processes that increase either the amount or quality of the gasoline. Gasolines contain hundreds of individual hydrocarbons which range from normal butane to C_{11} hydrocarbons such as methyl naphthalene. The properties of commercial gasolines are predominantly influenced by the refinery practices employed and partially influenced by the nature of the crude oils from which they are produced. Finished gasolines boil from about 30 to 225°C (85 to 437°F).

Gasolines are blended to satisfy diverse automotive requirements. Antiknock rating, distillation characteristics, vapor pressure, sulfur content, oxidation stability, and anticorrosion behavior are balanced to provide satisfactory vehicle performance. Additives may be used to provide or enhance specific performance features.

This chapter summarizes the significance of the more important physical and chemical characteristics of automotive gasolines and describes pertinent test methods for defining or evaluating these properties.

GRADES OF GASOLINES

For many years, two grades of automotive gasolines were marketed in the United States—"premium" and "regular"—and octane number was the main defining property. Some gasolines were also marketed as "super-premium," "intermediate," and "sub-regular" grades and as blends of high- and low-octane grades.

Until 1970, with the exception of one premium brand gasoline marketed on the East Coast and Southern areas of the United States, all grades of automotive gasolines contained lead alkyl compounds to increase octane rating. The average Antiknock Index (that is, average of Research Octane Number and Motor Octane Number) of the leaded premium grade increased steadily from about 82 at the end of World War II to about 96 in 1968. During the same time, the Antiknock Index of the leaded regular grade followed a parallel trend from about 77 to 90. Since 1973, the average Antiknock Index of the premium leaded grade has declined to about 92 while that of the leaded regular grade has declined to about 89. The premium leaded grade is no longer marketed in most of the United States but is still sold in the West Coast states and Hawaii.

In 1971, U.S. passenger car manufacturers began a transition to engines that would operate satisfactorily on gasolines with lower octane ratings, namely, a minimum Research Octane Number (RON) of 91. This octane level was chosen because unleaded gasolines are needed to prolong the effectiveness of automotive emission catalyst systems and because unleaded gasolines of 91 RON could be produced in the required quantities using refinery processing equipment then available. In 1970, gasoline marketers introduced unleaded and low-leaded gasolines of this octane level to supplement the conventional gasolines already available.

Beginning in July 1974, the U.S. Environmental Protection Agency (EPA) mandated that most service stations have available a grade of unleaded gasoline defined as having a lead content not exceeding 0.05 g of lead per U.S. gallon (0.05 g Pb/gal) and a Research Octane Number of at least 91. (This was changed to a minimum Antiknock Index of 87 in 1983.) Also starting in 1975, most gasoline-powered automobiles and light trucks required the use of unleaded gasoline. With this requirement, low-lead gasolines (0.5 g Pb/gal) disappeared. In addition, leaded premium began to be superseded by unleaded premium in the late 1970s and early 1980s.

The Environmental Protection Agency has continued to mandate the reduction of lead usage by requiring an average lead usage by a refinery not to exceed 0.8 g Pb/gal starting in October 1979 and reducing to 0.5 g Pb/gal in October 1980 averaged over the total gasoline pool. In October 1982, the EPA changed the regulation restricting the amount of lead in leaded gasoline to 1.1 g Pb/gal. This was further reduced to 0.5 g Pb/gal effective 1 July 1985 and to 0.1 g Pb/gal effective 1 Jan. 1986 averaged for quarterly leaded gasoline production. It is anticipated that lead usage in U.S. motor gasolines may be banned entirely as early as 1 Jan. 1988.

ANTIKNOCK PERFORMANCE

The definitions and test methods for antiknock performance for automotive gasolines are set forth in ASTM Specification for Automotive Gasoline (D 439). The antiknock rating of gasoline is an important characteristic. If antiknock rating is too low, knock occurs. Knock is a high-pitch, metallic rapping noise. Gasoline with antiknock rating higher than that required for knock-free operation does not improve performance. However, vehicles equipped with knock limiters may show a performance improvement as the antiknock quality of the gasoline is increased; conversely, antiknock rating decreases may cause vehicle performance loss. The loss of power and the damage to an automotive engine due to knocking are generally not significant until the knock intensity becomes very severe. Heavy and prolonged knocking may cause power loss and damage to the engine.

Knock depends on complex physical and chemical phenomena highly interrelated with engine design and operating conditions. It has not been possible to completely characterize the antiknock performance of gasoline with any single measurement. The antiknock performance of a gasoline is related intimately to the engine in which it is used and the engine operating conditions. Furthermore, this relationship varies from one engine design to another and may even be different among engines of the same design due to normal production variations.

The antiknock rating of a gasoline is measured by several methods. These procedures employ single-cylinder laboratory engines and more realistic, but less precise, multicylinder engines in cars. If a large enough group of cars is tested, the precision may be as good as that from the single-cylinder engine tests. Two single-cylinder methods have been standardized to measure antiknock rating: ASTM Test Method for Knock Characteristics of Motor Fuels by the Research Method (D 2699/IP 237) and ASTM Test Method for Knock Characteristics of Motor and Aviation Fuels by the Motor Method (D 2700/IP 236). Another method which is used for quality control in gasoline blending is ASTM Test Method for Research and Motor Method Octane Ratings using On-line Analyzers (D 2885).

These single-cylinder engine test procedures employ a variable-compression-ratio engine. The Motor method operates at a higher speed and inlet mixture temperature than the Research method. The procedures relate the knocking characteristics of a test gasoline to standard fuels which are blends of two pure hydrocarbons—*iso*-octane (2,2,4-trimethylpentane) and *n*-heptane. These blends are called primary reference fuels. By definition, the octane number of *iso*octane is 100, and the octane number of *n*-heptane is zero. At octane levels below 100, the octane number of a given gasoline is the percentage by volume of *iso*-octane in a blend with *n*-heptane that knocks with the same intensity at the same compression ratio as the gasoline when compared by one of the standardized engine test methods. The octane number of a gasoline greater than 100 is based upon the required milliliters of tetraethyl lead to be added to *iso*octane to produce knock with the same intensity as the gasoline. The number of milliliters of tetraethyl lead in *iso*octane is converted to octane numbers greater than 100 by use of tables included in the Research and Motor methods referred to previously.

The octane number of a given blend of either *iso*octane and *n*-heptane or tetraethyl lead in *iso*octane is, by definition, the same for the Research and Motor methods. However, the Research and Motor octane numbers will be rarely the same for commercial gasolines. Therefore, when considering the octane number of a given gasoline, it is necessary to know the engine test method. Research octane number (RON) is, in general, the better indicator of antiknock rating for engines operating at full throttle and low-engine speed. Motor octane number (MON) is the better indicator at full throttle, high-engine speed, and part throttle, low- and high-engine speed. The difference between RON and MON is called "sensitivity." According to recent surveys of U.S. commercial gasolines, MON of the leaded and unleaded regular grades is about 8 units lower than RON, and for unleaded premium grade sensitivity is about 9 units.

The antiknock performance of gasolines in cars, or Road octane number (RdON), may be determined directly by using primary reference fuels and manually controlled ignition timing to vary the knocking tendency of the engine. The most commonly used test methods are the Coordinating Research Council, Inc. (CRC) Modified Borderline (F-27) and Modified Uniontown (F-28) research techniques. When using these procedures, the knocking tendencies of test gasolines and reference fuels usually are compared at the lowest audible level of knock. Road octane number of a gasoline is defined as being equal to the octane number of the primary reference fuel blend that produces the same knock intensity while operating under the specified test conditions.

For most automotive engines and operating conditions, the Road octane number of a gasoline will be between its RON and MON. The difference between RON and RdON is called "depreciation."

The relationship between RON and MON used to predict the road antiknock performance of gasolines is dependent upon the vehicles and operating conditions. Antiknock Index is a currently accepted method of relating RON and MON to actual road antiknock performance of cars. The Antiknock Index currently used in ASTM D 439 is the average of the RON and the MON—that is, (RON + MON)/2—and is frequently abbreviated as (R+M)/2. U.S. Regulations now require a label on each service station pump showing the minimum (R+M)/2 value of the gasoline dispensed. ASTM D 439 describes the Antiknock Index of the various levels currently available as shown in Table 1.

VOLATILITY

The volatility characteristics of a gasoline are of prime importance to the driveability of vehicles under all conditions encountered in normal service. The large variations in operating conditions and wide ranges of atmospheric temperatures and pressures impose many limitations on a gasoline if it is to give satisfactory vehicle performance. Gasoline that vaporizes too readily in pumps, fuel lines, and carburetors will cause decreased fuel flow to the engine, resulting in rough engine operation or stoppage (vapor lock). Conversely, gasolines that do not vaporize readily enough may cause hard starting and poor warm-up and acceleration, as well as unequal distribution of fuel to the individual cylinders, which may cause knock. Fuels that vaporize too readily also may cause, if certain atmospheric conditions exist, ice formation in the carburetor throat, resulting in rough idle and stalling.

The volatility of automotive gasoline must be carefully "balanced" to provide the optimum compromise among performance features that depend upon the vaporization behavior. Superior performance in one respect may give serious

TABLE 1. Gasoline antiknock indexes and their application.

Leaded Gasoline (For Vehicles Which Can or Must Use Leaded Gasoline)	
Antiknock Index (RON + MON)/2, min[a,b]	Application
87	Meets antiknock requirements of most 1971 and later model vehicles that can use leaded gasoline and of pre-1971 vehicles with low antiknock requirements.
88	Meets antiknock requirements of most 1970 and prior model vehicles that were designed to operate on leaded gasoline, and of 1971 and later model vehicles that can use leaded gasoline and have high antiknock requirements.
89	Meets antiknock requirements of medium and heavy duty trucks that require higher octane leaded gasoline.
92	Suitable for most vehicles with very high antiknock requirements that can use leaded gasoline.
Unleaded Gasoline (for Vehicles That Can or Must Use Unleaded Gasoline)	
Antiknock Index (RON + MON)/2, min[a,b]	Application
85	For vehicles with low antiknock requirements.
87[c]	Meets antiknock requirements of most 1971 and later model vehicles.
90	For most 1971 and later model vehicles with high antiknock requirements.

[a]Reductions for seasonal variations are allowed in accordance with Fig. 1 of Specification D 439.
[b]Reductions for altitude are allowed in accordance with Fig. 2 of Specification D 439.
[c]In addition, Motor octane number must not be less than 82.0.

trouble in another. Therefore, volatility characteristics of automotive gasolines must be adjusted for seasonal variations in atmospheric temperatures and geographical variations in altitude.

Gasolines are mixtures of many hydrocarbons and have a range of vapor pressures and boiling points. Vapor pressure of gasoline, when measured at 100°F (38°C) in a bomb having a 4/1 ratio of air to liquid, is known as Reid Vapor Pressure (RVP). It is described in ASTM Test Method for Vapor Pressure of Petroleum Products (Reid Method) (D 323). Another method that is used is ASTM Test Method for Vapor Pressure of Petroleum Products (Micromethod) (D 2551). Results obtained by D 2551 can be converted to Reid Vapor Pressure through the use of a correlation.

The tendency of a gasoline to vaporize is also characterized by determining a series of temperatures at which various percentages of the gasoline have evaporated as described in ASTM Method for Distillation of Petroleum Products (D 86/IP 123). The temperatures at which 10, 50, and 90 percent evaporation occurs are used to characterize the volatility of automotive gasolines. Another method that can be used to characterize the distillation characteristics is ASTM Test Method for Boiling Range Distribution of Gasoline and Gasoline Fractions by Gas Chromatography (D 3710).

Gasoline vaporization tendency may also be expressed in terms of vapor/liquid ratio (V/L) at temperatures approximating those found in critical parts of the fuel system as presented in ASTM Test Method for Vapor-Liquid Ratio of Gasoline (D 2533).

In general terms, the following relationships between volatility and performance apply:

(*a*) High vapor pressures and low temperatures for 10 percent evaporated are both conducive to ease of cold starting. However, under hot-operating conditions, they are also conducive to vapor lock and increased vapor formation in fuel tanks and carburetors. The amount of vapors formed in fuel tanks and carburetors, which must be contained by the evaporative loss control system, are related to the RVP and distillation temperatures. Thus, a proper balance must be maintained and seasonally adjusted for good overall performance.

(*b*) Although vapor pressure is a factor in the amount of vapor formation under vapor-locking conditions, vapor pressure alone is not a good index. A better index for measuring the vapor-locking performance of gasolines in current model cars is the temperature at which the gasoline generates 20 vapor/liquid ratio at atmospheric pressure. The lower the temperature at $V/L = 20$, the greater the tendency to cause vapor lock.

(*c*) The temperature for 50 percent evaporated is a broad indicator of warm-up and acceleration performance under cold-starting conditions. The lower the 50 percent evaporated temperature, the better the performance. However, carburetor icing tendency and consequent engine stalling are increased by low 50 percent evaporated temperatures. The temperatures for 10 and 90 percent evaporated are also indicators of warm-up performance under cold-starting conditions, but to a lesser degree than the 50 percent evaporated temperature.

(*d*) The temperature for 90 percent evaporated and the final boiling point (FBP) or end point indicate the amount of relatively high-boiling components in gasoline. A high 90 percent evaporated temperature, because it usually is associated with higher density and high-octane number components, may contribute to improved fuel economy and resistance to knock. If the 90 percent evaporated temperature and the FBP are too high, they can cause poor mixture distribution in the intake manifold and combustion chambers, excessive combustion chamber deposits, and excessive varnish and sludge deposits in the engine.

Volatility characteristics of automotive gasolines are defined by ASTM Specification for Automotive Gasoline (D 439). The ASTM specification provides five volatility classes which varying limits of maximum or minimum vaporization tendency to adjust for seasonal and geographical changes in temperature and the altitude in which the gasoline is to be used. These volatility characteristics have been established on the basis of broad experience and cooperation between gasoline suppliers and manufacturers and users of automotive vehicles and equipment. Gasolines meeting this specification usually have provided satisfactory performance in typical passenger car service. Obviously, certain equipment or operating conditions may require or permit variations from these limits. Also, inasmuch as each volatility class covers a relatively broad range of use temperatures, it is evident that these class limits may be used to formulate alternate intermediate limits for specific temperature levels or to adjust for alternate gasoline distribution practices.

OTHER PROPERTIES

In addition to providing acceptable antiknock performance and volatility characteristics, automotive gasolines must also provide for satisfactory engine cleanliness. The following properties have a direct bearing on the overall performance of a gasoline.

Workmanship and Contamination

A finished gasoline is expected to be visually free of undissolved water, sediment, and suspended matter. It should be clear and bright when observed at 21°C (70°F). Physical contamination may occur during distribution of the fuel. Control of such contamination is a matter requiring constant vigilance by refiners, distributors, and marketers. Solid and liquid contamination can lead to restriction of fuel metering orifices, corrosion, fuel line freezing, gel formation, and filter plugging. ASTM Test Method for Water and Sediment in Distillate Fuel by Centrifuge (D 2709), or ASTM Test Methods for Particulate Contaminant in Aviation Turbine Fuels (D 2276/IP 216), may be used to determine the presence of contaminants.

Lead Content

As mentioned earlier, almost all gasolines formerly contained lead alkyl compounds to improve antiknock rating. Constraints imposed by emission control regulations have led to the wide scale availability of unleaded gasolines and the restrictions of lead content in leaded gasolines. Although lead may not be added deliberately to gasolines marketed as unleaded gasolines, some contamination by small amounts of lead may occur in the distribution system.

The following methods have been standardized for determining the concentration of lead in gasolines:

1. ASTM Test Method for Lead in Gasoline, Volumetric Chromate Method (D 2547).
2. ASTM Test Method for Lead in Gasoline by X-Ray Spectrometry (D 2599).
3. ASTM Test Method for Trace Amounts of Lead in Gasoline (D 3116).
4. ASTM Test Method for Low Levels of Lead in Gasoline by X-Ray Spectrometry (D 3229).
5. ASTM Test Method for Lead in Gasoline by Atomic Absorption Spectrometry (D 3237).
6. ASTM Test Method for Lead in Gasoline—Iodine Monochloride Method (D 3341).
7. ASTM Method for Rapid Field Test for Trace Lead in Unleaded Gasoline (Colorimetric Method) (D 3348).

The concentration ranges over which these methods have been standardized are listed in Table 2.

Phosphorus Content

Organometallic compounds of phosphorus have been added frequently to leaded gasolines as combustion chamber deposit modifiers. However, since phosphorus adversely effects exhaust emission control system components, its concentration in unleaded gasolines has been severely restricted by the U.S. Environmental Protection Agency. Its concentration can be determined by ASTM Test Method for Phosphorus in Gasoline (D 3231).

Manganese Content

Organomanganese compounds such as methyl cyclopentadienyl manganese tricarbonyl (MMT) have been added to some gasolines for octane improvement. However, the use of MMT in unleaded gasolines was banned by EPA in the United States in October 1978 since it results in higher emissions of hydrocarbons in auto exhaust. MMT is being used in unleaded gasolines in Canada and has had limited use at low dosages as a supplemental octane booster or trimming agent in some leaded gasolines in the United States. The manganese content can be determined by ASTM Test Method for Manganese in Gasoline by Atomic Absorption Spectrometry (D 3831).

Sulfur Content

Crude petroleum contains sulfur compounds, most of which are removed during refining. Currently, the average sulfur content of gasolines distributed in the United States is about 0.03 to 0.04 percent by weight, with the maximum reported values generally being less than 0.15 percent by weight.

Sulfur oxides formed during combustion may be converted to acids that promote rusting and corrosion of engine parts and exhaust systems. Sulfur oxides formed in the exhaust are undesirable atmospheric pollutants; however, the contribution of automotive exhaust to total sulfur oxide emissions is negligible. Sulfur content is usually determined by ASTM Test Method for Sulfur in Petroleum Products (Lamp Method) (D 1266/IP 107) or by ASTM Test Method for Sulfur in Petroleum Products

TABLE 2. ASTM methods for determining various concentrations of lead in gasoline.

ASTM Test Method	Concentration of Lead		
	g/U.S. gal	g/U.K. gal	g/L
D 2547	0.2 to 4.2	0.24 to 5.0	0.05 to 1.1
D 2599	0.1 to 5.0	0.12 to 6.00	0.026 to 1.32
D 3116	0.001 to 0.1	0.0012 to 0.12	0.00026 to 0.026
D 3229	0.010 to 0.50	0.012 to 0.60	0.0026 to 0.132
D 3237	0.010 to 0.10	0.012 to 0.12	0.0025 to 0.025
D 3341	0.1 to 5.0	0.12 to 6.0	0.026 to 1.3
D 3348	0.01 to 0.10	0.012 to 0.12	0.00264 to 0.0264

(X-Ray Spectrographic Method) (D 2622). It may also be determined by ASTM Test Method for Trace Quantities of Sulfur in Light Liquid Hydrocarbons by Oxidative Microcoulometry (D 3120). The presence of free sulfur or reactive sulfur compounds can be detected by ASTM Method for Detection of Copper Corrosion from Petroleum Products by the Copper Strip Tarnish Test (D 130/IP 154). Sulfur in the form of mercaptans can be determined by ASTM Test Method for Mercaptan Sulfur in Gasoline, Kerosine, Aviation Turbine, and Distillate Fuels (Potentiometric Method) (D 3227).

Existent Gum and Stability

During storage, gasolines may oxidize slowly in the presence of air and form undesirable oxidation products called gum. The gum is usually soluble in the gasoline but may appear as a sticky residue on evaporation. These residues may deposit on carburetor surfaces, intake manifold, intake valves, stems, guides, and ports. ASTM Specification for Automotive Gasoline (D 439) limits gasoline to a maximum of 5 mg of existent gum/100 ml of gasoline. ASTM Test Method for Existent Gum in Fuels by Jet Evaporation (D 381/IP 131) is a test to determine the amount of existing gum. Many motor gasolines are deliberately blended with nonvolatile oils or additives or both, which remain as residues in the evaporation step of the existent gum test. A heptane washing step is therefore a necessary part of the procedure to remove such materials so that the deleterious material or existent gum may be determined.

Automotive gasolines have a negligible gum content when manufactured, but may form gum during extended storage. ASTM Test Method for Oxidation Stability of Gasoline (Induction Period Method) (D 525/IP 40) is a test to indicate the tendency of a gasoline to form gum in storage. It should be recognized, however, that the method's correlation may vary markedly under different storage conditions and with different gasoline blends. Most automotive gasolines contain special chemicals (antioxidants) to prevent oxidation and gum formation. Some gasolines also contain metal deactivators for this purpose. Commercial gasolines available in service stations move rather rapidly from refinery production to vehicle usage and are not designed for extended storage. Gasolines purchased for severe bulk storage conditions or for prolonged storage in vehicle fuel systems generally have additional amounts of antioxidant and metal deactivator added.

Gravity

"Gravity" is a term used to denote the density of gasolines. Two methods of expressing gravity commonly are used. Specific gravity or relative density is the ratio of the mass of a given volume of gasoline at a temperature, usually specified at 15.6°C (60°F), to the mass of an equal volume of water at the same temperature. Typically, automotive gasolines have specific gravities between 0.71 and 0.79 and American Petroleum Institute (API) gravities between 69 and 48. API gravity is based on an arbitrary hydrometer scale and is related to specific gravity as follows:

API gravity, deg =

$$\frac{141.5}{\text{sp gr } (15.6/15.6°\text{C})} - 131.5$$

Gravity of gasoline is determined using ASTM Test Method for Density, Relative Density (Specific Gravity), or API Gravity of Crude Petroleum and Liquid Petroleum Products by Hydrometer Method (D 1298/IP 160).

Rust and Corrosion

Filter plugging and engine wear problems are reduced by minimizing rust and corrosion in fuel distribution and vehicle fuel systems. Modifications of ASTM Test Method for Rust-Preventing Characteristics of Inhibited Mineral Oil in the Presence of Water (D 665/IP 135) are sometimes used to measure rust protection of gasolines.

Hydrocarbon Composition

The major hydrocarbon-type constituents of gasoline are paraffins, olefins, and aromatics, and they are identified by using ASTM Test Method for Hydrocarbon Types in Liquid Petroleum Products by Fluorescent Indicator Adsorption (D 1319/IP 156). The amount of benzene can be determined by ASTM Test Method for Benzene in Motor and Aviation Gasoline by Infrared Spectroscopy (D 4053). The amounts of benzene and other aromatics can be determined by ASTM Test Method for Benzene and Toluene in Finished Motor and Aviation Gasoline by Gas Chromatography (D 3606) and by ASTM Test Method for Aromatics in Finished Gasoline by Gas Chromatography (D 4420). A variety of other nonstandardized gas chromatographic procedures are also available to permit more detailed analyses.

Additives

Gasoline additives are used to enhance or provide various performance features related to the satisfactory operation of engines as well as to minimize gasoline handling and storage problems. These compounds complement refinery processing in attaining the desired level of product quality.

The amount and variety of additives used in gasolines have grown rapidly. For many years, antiknock compounds were almost the only additives used; however, now the list also includes combustion chamber deposit modifiers, antioxidants, metal deactivators, corrosion or rust inhibitors, carburetor antiicing additives, fuel line antifreeze agents, gasoline detergents, gasoline dispersants, and identifying dyes.

Gasoline additives currently available are shown by class and function in Table 3.

U.S. LEGAL RESTRICTIONS ON BLENDING ADDITIVES AND ALTERNATIVE FUELS INTO GASOLINE

The U.S. Environmental Protection Agency (EPA) has established vehicle exhaust emission and fuel evaporative emission standards as part of the U.S. national effort to attain acceptable ambient air quality. To meet these EPA vehicle requirements, extensive modifications have been made to automotive engines and emission systems. Since some fuel components can harm the effectiveness of vehicle emission control systems, EPA also exercises control over automotive fuels. EPA regulations on availability of unleaded gasolines, and on limits of lead, phosphorus, and manganese contents in gasolines, have been mentioned above.

In addition, current laws and regulations prohibit the introduction into U.S. commerce, or increases in concentration in use of, any fuel or fuel additive for general use for post 1974 model-year light duty motor vehicles, which is not "substantially similar" to the fuel or fuel additive utilized in the emission certification of such vehicles.

Fuels are considered by EPA to be "substantially similar" if the following criteria are met:

1. The fuel must contain carbon, hydrogen, and oxygen, nitrogen, and/or sulfur, exclusively in the form of some combination of the following:

a. Hydrocarbons.

b. Aliphatic ethers.

c. Aliphatic alcohols other than methanol.

d. (i) Up to 0.3 percent methanol by volume; (ii) up to 2.75 percent by volume with an equal volume of butanol, or higher molecular weight alcohol.

e. A fuel additive at a concentration of no more than 0.25 percent by weight which contributes no more than 15 ppm sulfur by weight to the fuel.

2. The fuel must contain no more than 2.0 percent oxygen by weight.

3. The fuel must possess, at the time of manufacture, all of the physical and chemical characteristics of an unleaded gasoline as specified by ASTM Standard D 439 (or applicable emergency standard if one has been instituted) for at least one of the Seasonal and Geographical Volatility Classes specified in the standard.

4. The fuel additive must contain only carbon, hydrogen, and any or all of the following elements; oxygen, nitrogen, and sulfur.

Fuels or fuel additives which are "not substantially similar" may only be used if a waiver of this prohibition is obtained from EPA. Manufacturers of fuels and fuel additives must apply for such a waiver and must establish to the satisfaction of EPA that the fuel or additive does not cause or

TABLE 3. Commercial gasoline additives.

Class or Function	Common Additive Type
1. Antiknock compounds—to improve research, motor, and road octane quality.	lead alkyls, such as tetraethyl lead, tetramethyl lead, and physical or reacted mixtures: organomanganese compounds, such as methyl cyclopentadienyl manganese tricarbonyl
2. Combustion deposit modifiers—to minimize surface, ignition, rumble, preignition, and spark plug fouling.	organic or organometallic compounds usually containing phosphorus
3. Antioxidants—to minimize oxidation and gum formation in gasoline and improve handling and storage characteristics.	phenylenediamine and alkyl phenol compounds
4. Metal deactivators—to deactivate small traces of copper and other metal ions that are powerful oxidation catalysts.	N, N′-disalicylidene-1,2,-propane diamine
5. Corrosion or rust inhibitors—to minimize corrosion and rusting of fuel system and storage and handling facilities.	organic acids, amine salts, and derivatives of carboxylic acids, many of which have surface-active properties
6. Carburetor antiicing additives—to minimize engine stalling due to ice accumulation on the throttle.	derivatives of carboxylic acids that have surface-active properties: freeze point depressants such as alcohols and glycols
7. Gasoline detergents—to remove or minimize the accumulation of deposits in the throttle section of the carburetor, which adversely affect the metering characteristics.	amines and derivatives of caboxylic having surface-active properties, some of which are polymeric
8. Gasoline dispersants—to minimize the accumulation of deposits in the carburetor, intake manifold, intake ports, and underside of the intake valves.	derivatives, usually aminoimido, of polymeric hydrocarbons, occasionally used in combination with special oils
9. Deposit control additives—to remove or minimize the accumulation of deposits throughout the carburetor, intake manifold, intake ports, and underside of the intake valves.	derivatives, usually amino of polymeric hydrocarbons of molecular weight in the 1000 to 2000 range: often used in combination with special oils or polyolefins
10. Dyes—to identify gasoline blends.	oil soluble and liquid dyes

NOTE: Some materials may also be marketed as multifunctional or multipurpose additives performing more than one of the functions described in Items 5, 6, 7, 8, and 9.

contribute to a failure of any emission control device or system over the useful life of the vehicle for which it was certified. If the EPA Administrator has not acted to grant or deny the waiver application within 180 days after its receipt, the waiver is treated as granted. Thus far, EPA has granted several waivers for blends of oxygenates with gasolines. The reader is referred to EPA for the latest information on waivers and the conditions under which they may be used.

GASOLINE-OXYGENATE BLENDS

Gasolines are not the only fuels used in internal combustion spark ignition engines. Blends of gasolines with oxygenates are already common in the U.S. marketplace under terms of EPA waivers. These blends consist primarily of gasoline with a substantial amount of oxygenates, which are oxygen-containing ashless, organic compounds such as alcohol or ether. Some of the test methods cited above are not applicable to such gasoline-oxygenate blends. Also, new test methods and limits need to be developed to protect against phase separation, incompatibility with elastomers and plastics, corrosion of metals, and other factors that may affect vehicle performance and durability. ASTM is developing a specification that encompasses all fuels for automotive spark-ignition engines. It appears in Volume 05.01 of the 1987 Annual Book of ASTM Standards as D-2 Proposal P 176 Proposed Specification for Automotive Spark-Ignition Engine Fuel.

Oxygenates which have been used commonly to date include methanol, ethanol, tertiary butyl alcohol (TBA), and methyl tertiary butyl ether (MTBE). When methanol is used, cosolvents must be used to help prevent separation of the blend into two phases in the presence of water. Such a blend is known as a gasoline-alcohol blend.

Recognizing that specifications for fuel grade oxygenates may be needed, ASTM is already developing a specification for fuel grade ethanol to be used in blends. It appears in Volume 05.01 of the 1987 Annual Book of ASTM Standards as D-2/E-44 Proposal P 170 Proposed Specification for Denatured Fuel Ethanol to be Blended with Gasoline for Use as an Automotive Spark-Ignition Engine Fuel.

Sampling of Gasoline-Oxygenate Blends

Sampling of blends can be conducted according to ASTM Practice for Manual Sampling of Petroleum and Petroleum Products (D 4057) except that the water displacement (11.3.1.8 of Practice D 4057) must not be used.

Test Methods in ASTM D 439 that Apply to Gasoline-Oxygenate Blends

Some of the test methods cited above for gasolines can be used without modification to measure the properties of gasoline-oxygenate blends. These properties include:

1. Distillation temperatures.
2. Lead content.
3. Sulfur content.
4. Copper corrosion.
5. Existent gum.
6. Oxidation stability.

The specification limits in D-2 Proposal P 176 for these properties are identical to those in ASTM D 439 for gasolines.

Test Methods Requiring Modification for Gasoline-Oxygenate Blends

When oxygenates are present, some properties must be measured by modifications of test procedures cited in ASTM D 439 or by additional test procedures under development. It is therefore necessary to determine whether oxygenates are present in order to select the appropriate test methods.

At the present time no simple laboratory test is available to detect the presence of and to determine the amount of various oxygenates which may be present in gasolines. However, D-2 Proposal P 176 contains a water-extraction method which can detect the presence of the commonly used low molecular weight alcohols, methanol through butanols, but does not detect ethers. In this method, a sample of the gasoline is shaken with distilled water to extract the water-soluble alcohols. After separating into two phases, the lower aqueous phase is placed on a refractometer. A minimum increase in refractivity of light above that of distilled water indicates that alcohol(s) is present in a sufficient amount that, in certain cases, methods other than those in ASTM D 439 must be used.

The refractometer method does not distinguish between the types and amounts of alcohols and may not detect the presence of other oxygenates which are not readily extracted into water, for example, MTBE. Therefore, more sophisticated gas chromatographic (GC) methods described in D-2 Proposal P 176 are under development and standardization. They will detect the presence and concentrations of individual oxygenates. Such GC methods must be used to avoid adding an oxygenate to a blend which already contains some type of oxygenate and to ensure compliance with EPA rules and waivers.

When oxygenates are present in a fuel, ASTM Test Method for Vapor Pressure of Petroleum Products (Reid Method) (D 323) and ASTM Test Method for Vapor-Liquid Ratio of Gasoline (D 2533) cannot be used without modification. In addition, ASTM Test Method for Knock Characteristics of Motor Fuels by the Research Method (D 2699) and ASTM Test Method for Knock Characteristics of Motor and Aviation Fuels by the Motor Method (D 2700) may need special precautions or some modifications.

Vapor Pressure

Test Method D 323 (Reid Method) cannot be used with gasoline-oxygenate blends which contain water-extractable oxygenates because the fuel sample comes into contact with water. Therefore D-2 Proposal P 176 includes two proposed replacement methods, namely a dry method using the Reid apparatus and an automatic vapor pressure method which can also be used with gasolines as well as with blends.

The dry method uses the same apparatus and essentially the same procedure as the Reid Method (D 323) except that precautions are taken to keep water out of the apparatus.

The automatic method utilizes a special automatic Reid vapor pressure instrument. The cold sample cup of the instrument is filled with the chilled, air-saturated fuel sample and connected to the instrument. The instrument operation is started and the vapor pressure is determined automatically.

ASTM Test Method for Vapor Pressure of Petroleum Products (Micromethod) (D 2551) can be used without modification for gasoline-oxygenate blends. However, its future is questionable because the apparatus is no longer readily available and special precautions are necessary to avoid exposure to toxic vapors of mercury.

It should be noted that the vapor pressure limits for gasolines and for gasoline-oxygenate blends in D-2 Proposal P 176 are identical to those for gasolines in ASTM Specification D 439.

Vapor-Liquid Ratio

ASTM Test Method D 2533 is not applicable to fuels containing alcohols, ethers, or other compounds soluble in glycerin. Two proposed procedures are therefore included in D-2 Proposal P 176. One is a modification of ASTM Test Method D 2533 in which the glycerin is replaced by mercury. This method is included for information only because careful handling is required to avoid exposure to toxic vapors of mercury. The other one is a modification of the dry vapor pressure method in which the volume of the two-piece bomb is 21 times (instead of 5 times) the volume of the fuel sample used. In this latter method, a vapor-liquid ratio of 20 is calculated from a pressure increase.

The Appendix of ASTM Specification D 439 includes a computer method, a linear equation method, and a nomograph method that can be used for estimating vapor-liquid ratio of gasolines from Reid vapor pressure and distillation test results. However, these estimation methods are not applicable to gasoline-oxygenate blends.

D-2 Proposal P 176 does not contain any limits on temperature for a vapor/liquid ratio for gasoline-oxygenate blends because an acceptable test method has not yet been standardized and because no correlation with automotive service has yet been demonstrated.

Octane Number

The addition of oxygenates to gasolines alters the stoichiometry, latent heat of vaporization, and volatility characteristics and may also effect the nature and amount of combustion chamber deposits. The resultant, combined effects on octane number determination are not well established. Accordingly, ASTM Subcommittee D02.01 on Combustion Characteristics is undertaking an investigation to quantify the effects on octane number determinations using oxygenates in fuels and will modify Test Method D 2699 and D 2700 as necessary.

New Test Methods Required for Gasoline-Oxygenate Blends

Appropriate test methods must be developed and standardized and limits must be set on such properties of blends as water tolerance, compatibility with plastics and elastomers, and metal corrosion.

Water Tolerance

Gasolines and water are almost entirely immiscible; when mixed, they readily separate into two phases. Blends will dissolve some water but will also separate into two phases when contacted with more water than they can dissolve. This water may be absorbed from ambient air or may occur as liquid water in the bottom of tanks in the storage, distribution, and vehicle fuel system. Such separation may not be a problem with gasoline-ether blends. However, it is a matter of concern with gasoline-alcohol blends because low molecular weight alcohols are readily extracted from blends leaving an alcohol-poor gasoline phase and an alcohol-rich aqueous phase. This aqueous phase may be highly corrosive to many metals and the engine will not operate on it. In addition, the properties of the gasoline phase will differ from those of the blend, for example, octane rating, volatility, stoichiometry, etc., and engine performance may suffer.

The amount of water which can be dissolved in blends depends upon the nature and amount of alcohols used, the specific hydrocarbons present, and the temperature of the blend. Gasoline-oxygenate blends must therefore be tested at the lowest temperatures to which they may be subjected. Since temperatures vary geographically and seasonally, D-2 Proposal P 176 specifies the maximum temperature for phase separation for each month of the year for each of the 50 states or portion thereof in the United States.

At the present time, three different water solubility tests are being considered for standardization by ASTM.

Compatibility with Plastics and Elastomers

All plastics and elastomers used in current automotive fuel systems such as gaskets, "0"-rings, diaphragms, filters, seals, etc. may be affected in time by exposure to motor fuels. These effects include dimensional changes, embrittlement, softening, delamination, increase in permeability, loss of plasticizers, and disintegration. Gasoline-oxygenate blends may aggravate these effects.

The effects depend upon the type and amount of the oxygenates in the blend, the aromatic content of the gasoline, the generic polymer and specific composition of the elastomeric compound, the temperature and duration of contact, and whether the exposure is to liquid or vapor.

Currently there are no generally accepted tests that correlate with field experience to allow estimates of tolerance to oxygenates of specific plastics or elastomers.

Metal Corrosion

Corrosion of metals on prolonged contact is a problem with gasolines alone but is generally more severe with gasoline-oxygenate blends. The aqueous phase, which may separate on water contamination of blends of gasolines and low-molecular-weight alcohols, is particularly aggressive in its attack on the terne (lead-tin alloy) coat of fuel tanks. Aluminum and zinc castings of carburetors, and steel components such as fuel senders, fuel tubes, and pump housings, are also affected by the aqueous phase.

A number of test procedures, other than long-term vehicle tests, have been used or proposed to evaluate the corrosive effects of fuels on metals. The tests range from static soaking of metal coupons to operation of a complete automotive fuel system. None of these tests have yet achieved the status of an ASTM standard.

Applicable ASTM Specifications

D 439	Specification for Automotive Gasoline
D-2/E-44 Proposal P 170	Proposed Specification for Denatured Fuel to be Blended with Gasolines for Use as an Automotive Spark-Ignition Engine Fuel
D-2 Proposal P 176	Proposed Specification for Automotive Spark-Ignition Engine Fuel

Applicable ASTM/IP Standards

ASTM	*IP*	*Title*
D 86	123	Method for Distillation of Petroleum Products
D 130	154	Method for Detection of Copper Corrosion from Petroleum Products by the Copper Strip Tarnish Test
D 323		Test Method for Vapor Pressure of Petroleum Products (Reid Method)

Applicable ASTM/IP Standards ***(continued)***

ASTM	*IP*	*Title*
D 381	131	Test Method for Existent Gum in Fuels by Jet Evaporation
D 525	40	Test Method for Oxidation Stability of Gasoline (Induction Period Method)
D 665	135	Test Method for Rust-Preventing Characteristics of Inhibited Mineral Oil in the Presence of Water
D 1266	107	Test Method for Sulfur in Petroleum Products (Lamp Method)
D 1298	160	Test Method for Density, Relative Density (Specific Gravity), or API Gravity of Crude Petroleum and Liquid Petroleum Products by Hydrometer Method
D 1319	156	Test Method for Hydrocarbon Types in Liquid Petroleum Products by Fluorescent Indicator Adsorption
D 2276	216	Test Method for Particulate Contaminant in Aviation Turbine Fuels
D 2533		Test Method for Vapor-Liquid Ratio of Gasoline
D 2547	248	Test Method for Lead in Gasoline, Volumetric Chromate Method
D 2551		Test Method for Vapor Pressure of Petroleum Products (Micromethod)
D 2599	228	Test Method for Lead in Gasoline by X-Ray Spectrometry
D 2622		Test Method for Sulfur in Petroleum Products (X-Ray Spectrographic Method)
D 2699	237	Test Method for Knock Characteristics of Motor Fuels by the Research Method
D 2700	236	Test Method for Knock Characteristics of Motor and Aviation Fuels by the Motor Method
D 2709		Test Method for Water and Sediment in Distillate Fuels by Centrifuge
D 2885		Test Method for Research and Motor Method Octane Ratings Using On-Line Analyzers
D 3116		Test Method for Trace Amounts of Lead in Gasoline
D 3227	342	Test Method for Mercaptan Sulfur in Gasoline, Kerosine, Aviation Turbine, and Distillate Fuels (Potentiometric Method)
D 3229		Test Method for Low Levels of Lead in Gasoline by X-Ray Spectrometry
D 3231		Test Method for Phosphorus in Gasoline
D 3237		Test Method for Lead in Gasoline by Atomic Absorption Spectrometry
D 3341		Test Method for Lead in Gasoline—Iodine Monochloride Method
D 3348		Method for Rapid Field Test for Trace Lead in Unleaded Gasoline (Colorimetric Method)
D 3606		Test Method for Benzene and Toluene in Finished Motor and Aviation Gasoline by Gas Chromatography

Applicable ASTM/IP Standards ***(continued)***

ASTM	IP	Title
D 3710		Test Method for Boiling Range Distribution of Gasoline and Gasoline Fractions by Gas Chromatography
D 3831		Test Method for Manganese in Gasoline by Atomic Absorption Spectrometry
D 4057		Practice for Manual Sampling of Petroleum and Petroleum Products
D 4420		Test Method for Aromatics in Finished Gasoline by Gas Chromatography

Aviation Fuels

5

INTRODUCTION

IT IS DIFFICULT TO DISCUSS AVIATION FUELS without reviewing the development history of the various types of aviation fuels and describing quality requirements in terms of official specifications produced by the cooperative efforts of engine manufacturers, airline operators, fuel suppliers, and appropriate government departments. These documents define the required fuel properties and specify the standard test methods to be used. The international validity of these specifications and rigid enforcement ensures that fuels of uniform quality are available on a worldwide basis for all types of aircraft engines.

It is not feasible to include full details of all major international specifications in this chapter. Even summaries of the main requirements would be of little permanent value, since these specifications are revised and updated frequently to meet new aircraft needs or reflect changing supply situations. However, the basic content of the various specifications covering similar grades of fuel do not differ greatly, and, with few exceptions, the same fuel properties are controlled in each. Typical examples of the physical and chemical property requirements contained in current specifications are included for each of the main aviation gasoline and jet fuel grades.

HISTORICAL DEVELOPMENT OF AVIATION FUELS

Aviation gasolines for spark-ignition engines reached their development peak in the 1939 to 1945 war years. The advent of the gas turbine inhibited further piston engine development, and, although large quantities of aviation gasoline will be required for many years, quality requirements are unlikely to change significantly.

The first aviation gas-turbine engines were regarded as having noncritical fuel requirements. Since ordinary illuminating kerosine was the original development fuel, the first turbine fuel specifications were written largely around the properties and test methods associated with this well-established product. With increased complexity in design of the engine and its control, fuel specification tests have become inevitably more complicated and numerous. Current demands for improved performance, economy, and overhaul life will indirectly continue the trend towards additional tests; nevertheless, the optimum compromise between fuel quality and availability is achieved largely by the current fuel specifications.

AVIATION GASOLINE

Composition and Manufacture

Aviation gasoline is the most complex fuel produced in a refinery. Strict process control is required to ensure that the stringent (and sometimes conflicting) specifications are met for volatility, calorific value, and antiknock ratings. In addition, careful handling is required during storage and distribution to guard against various forms of contamination which can affect such properties as volatility, gum values, and the copper strip corrosion test.

Aviation gasoline consists substantially of hydrocarbons. Sulfur-containing and oxygen-containing impurities are limited strictly by specification and only certain additives are permitted (refer to the section on Aviation Fuel Additives).

The main component of high-grade aviation gasolines is *iso*octane produced in the alkylation process by reaction of refinery butenes with isobutane over acid catalysts. To meet volatility requirements for the final blend, a small proportion of isopentane (obtained by superfractionation of light straight-run gasoline) is added. The aromatic component required to improve rich mixture rating is usually a catalytic reformate. The amount of aromatic components added is limited indirectly by the gravimetric calorific value requirement.

Only grade 80 fuel can include a proportion of straight-run gasoline because straight-run gasolines, which contain varying amounts of paraffins, naphthenes, and aromatics invariably lack the necessary branch-chain paraffins (isoparaffins) required to produce the higher grade fuels.

Specifications

Content

Aviation fuel specifications generally contain three main sections covering suitability, composition, and chemical and physical requirements.

The suitability section is included as a safeguard against the possible failure in service of a fuel which meets all the published physical and chemical tests in the specification. It throws the onus on the fuel producer to obey the spirit as well as the letter of the law. This philosophy is inherent in all aviation fuel specifications.

The composition section stipulates that the fuel must consist entirely of hydrocarbons except for trace amounts of approved additives, such as alkyl lead antiknock additive, dyes, and oxidation inhibitors. Its main importance is in listing the approved additives and, indirectly, in excluding any nonhydrocarbon blending components such as oxygenates, which might be used to improve a critical property of the fuel at the ultimate expense of other fuel properties.

The chemical and physical requirements section is the one most familiar to users since it carefully defines the allowable limits for many chemical and physical properties of the fuel and the standard test methods to be employed.

Fuel Grades

About six basic fuel grades have been in use since the 1939 to 1945 war period. In recent years, the diminishing demand for aviation gasoline has led to a reduction in the number of grades available. With fewer fuel grades, manufacturing, storage, and handling costs were reduced with subsequent benefits to consumers. At present, three grades—80, 100, and 100 lowlead—are specified in ASTM Specification for Aviation Gasolines (D 910).

Specifications covering the various grades have been drawn up by a number of bodies, and these have been revised as engine requirements changed. The most commonly quoted aviation gasoline specifications are those issued by the U.S. Department of Defense (military specifications), the British Ministry of Defense (DERD[1] specifications), and the American Society for Testing and Materials (ASTM D 910). Table 1 lists the main aviation gasoline specifications in current use and indicates the various grades together with their identifying dye colors.

Due to the international nature of aviation activities, the technical requirements of all the Western specifications are virtually identical, and only differences of a minor nature exist between the specifications issued in the various major countries. The Soviet GOST specifications (and their East European equivalents) differ in the grades covered and also in respect to some of the limits applied, but, in general, the same fuel properties are controlled, and most test methods basically are similar to their Western equivalents [American Society for Testing and Materials (ASTM) and Institute of Petroleum (IP) standards]. Soviet aviation gasoline grades are summarized in Table 2.

Table 3 provides detailed requirements for aviation gasoline as contained in ASTM Specification for Aviation Gasolines (D 910). In general, the main technical requirements of all other Western specifica-

[1]In current issues of the British Military Specifications, the traditional term "D.Eng.R.D." has been abbreviated to "DERD" (Directorate of Engine Research and Development). For uniformity, this new abbreviation is used throughout this chapter, even for obsolete specifications.

TABLE 1. Aviation gasolines—main international specification grades.

Identifying Color	Nominal Antiknock Characteristics, Lean/Rich	NATO Code Number	Current Specifications: DERD 2485 British Ministry of Defense	Current Specifications: MIL-G-5572 U.S. Department of Defense[b]	Current Specifications: ASTM D 910	Use
Colorless	73	F-13[a]	...	...	...	blending component
Colorless	80	...	...	...	...	blending, historic
Red	80/87	F-12	80	80/87	80	minor civil
Blue	91/96	F-15[a]		obsolete		
Blue	100/130	F-18	100LL	100/130	100LL	major civil
Green	100/130	...	100	...	100	minor military
Brown	108/135			obsolete		
Purple	115/145	F-22	115	115/145	...	military—virtually obsolete

[a]Obsolete designation.
[b]Specification MIL-G-5572 was withdrawn in 1988.

TABLE 2. Soviet aviation gasoline grades.

Specification	Grade	Color	Use
...	B.70	colorless	obsolete
GOST-1012	B.91/115[a]	green	current
GOST-1012	B.95/130[a]	yellow	current
...	B.100/130	bright orange	obsolete
GOST-5760	BA(115/160)	varies	obsolete

[a]In regular and premium qualities.

tions are virtually identical to those in Table 3, although differences occur in the number of grades covered and, in some cases, the amount of tetraethyl lead (TEL) permitted. The various grades within the specification differ fundamentally in only a few vital respects, such as color, antiknock ratings, and TEL content. This is true of all the Western aviation gasoline specifications. The two remaining grades in the Soviet GOST specification are subdivided, somewhat curiously, into ordinary and premium qualities with differing limits for aromatics, olefins, sulfur, and acidity.

The limits specified for Western grades of aviation gasoline were, in most cases, dictated originally by military aircraft engine requirements. Since then, the performance requirements for civil and military aircraft engines have changed very little. However, improved fuel manufacturing techniques and the reduced demand for certain grades has allowed fuel suppliers to produce modified fuel grades more suited to market requirements. In some cases, the objective has been to offer a technically superior fuel; in other cases, the aim has been the reduction of production, storage, and handling costs by providing a fuel suitable for use in a wider range of engine types than was possible with the standard grades.

Characteristics and Requirements

Antiknock Properties

The various fuel grades are classified by their "antiknock" quality characteristics as determined in single-cylinder laboratory engines. Knock, or detonation, in an engine is a form of abnormal combustion where the air/fuel charge in the cylinder ignites spontaneously in a localized area instead of being consumed progressively by the spark-initiated flame front. Knocking combustion can damage the engine and give serious power loss if allowed to persist. The various grades are designed to guarantee knock-free operation for a range of engines from those used in light aircraft up to high-powered transport and military types.

TABLE 3. Detailed requirements for aviation gasolines.[a]

	Grade 80	Grade 100	Grade 100LL
Knock value, lean rating:			
Minimum octane number	80	100	100
Knock value, rich rating:			
Minimum octane number	87	...	...
Minimum performance number	...	130	130
Color	Red	Green	Blue
Dye content:			
Permissible blue dye, max, mg/U.S. gal	0.5	4.7	5.7
Permissible yellow dye, mg/U.S. gal	None	5.9	None
Permissible red dye, max, mg/U.S. gal	8.65	None	None
Tetraethyl lead, max, mL/U.S. gal	0.5	4.0	2.0
gPb/L	0.14	1.12	0.56
	Requirements for All Grades		
Distillation temperature, °C (°F):			
10% evaporated, max temp	75(167)		
40% evaporated, min temp	75(167)		
50% evaporated, max temp	105(221)		
90% evaporated, max temp	135(275)		
Final boiling point, max, °C (°F)	170(338)		
Sum of 10 and 50% evaporated temperatures, min, °C (°F)	135(307)		
Distillation recovery, min, %	97		
Distillation residue, max, %	1.5		
Distillation loss, max, %	1.5		
Net heat of combustion, min, Btu/lb (MJ/kg)	18720 (43.54)		
Vapor pressure:			
min, kPa(psi)	38(5.5)		
max, kPa(psi)	49(7.0)		
Copper strip corrosion, max	No. 1		
Potential gum (5-h aging gum), max, mg/100 mL	6		
Visible lead precipitate, max, mg/100 mL	3		
Sulfur, max %m	0.05		
Freezing point, max, °C(°F)	−58(−72)		
Water reaction	Volume change not to exceed ±2 mL		
Permissible antioxidants, max, lb/1000 bbl (42 gal)	4.2		

[a]ASTM Specification for Aviation Gasolines (D 910-85).

The antiknock ratings of aviation gasolines are determined in standard ASTM laboratory engines by matching their performance against reference blends of pure *iso*octane (2,2,4-trimethyl pentane) and n-heptane. Fuel rating is expressed as an octane number (ON) which is defined as the percentage of *iso*octane in the matching reference blend. Fuels of higher performance than pure *iso*octane (100 ON) are tested against blends of *iso*octane with various amounts of antiknock additive. The rating of such fuel is expressed as a performance number (PN) which is defined as the maximum knock-free power output obtained from the fuel expressed as a percentage of the power obtainable on *iso*octane.

The antiknock rating of fuel varies according to the air/fuel mixture strength employed. This fact is used in defining the performance requirements of the higher grade aviation fuels. As mixture strength is increased (richened), the additional fuel acts as an internal coolant and suppresses knocking combustion which, in turn, permits a higher power rating to be obtained. Since maximum power output is the prime requirement of an engine under rich take-off conditions, the "rich mixture performance" of a fuel is determined in a special supercharged single-cylinder engine using ASTM Test for Knock Characteristics of Aviation Fuels by the Supercharge Method (D 909/IP 119). Similarly, economic cruising operation of an engine is obtainable with weak (lean) mixture strengths. "Weak mixture performance" is determined by

ASTM Test for Knock Characteristics of Motor and Aviation Fuels by the Motor Method (D 2700/IP 236).

Until 1975, ASTM Specification for Aviation Gasolines (D 910) designated aviation gasoline grades with two numbers, for example, "grade 100/130." The lower number denoted an antiknock of 100 minimum by the lean mixture test procedure, and the higher number 130 minimum by the rich mixture procedure. Although the ASTM specification now uses only one number to designate grade (the number from the lean mixture procedure) some other specifications still use both.

Volatility

All internal combustion engine fuels must be easily convertible from storage in the liquid form to the vapor phase in the engine to allow formation of the combustible air/fuel vapor mixture. If gasoline fuel volatility is too low, liquid fuel enters the cylinders and washes lubricating oil from the walls and pistons. This would increase engine wear and cause dilution of the crankcase oil. Poor volatility can also give rise to critical maldistribution of mixture strength between cylinders. If volatility is too high, fuel can vaporize in the fuel tank and supply lines giving undue venting losses and the possibility of fuel starvation through "vapor lock" in the fuel lines. The cooling effect due to rapid vaporization of excessive amounts of highly volatile material also can cause ice formation in the carburetor under certain conditions of humidity and air temperature. Many modern aircraft have anti-icing devices on the engines including the provision of carburetor heating.

Distillation characteristics are determined with a procedure (ASTM D 86/IP 123) in which a sample of the fuel is distilled and the vapor temperature recorded for the percentages of evaporation or distillation throughout the range. Distillation points are selected to control volatility in the following ways:

1. The percent evaporated at 75°C (167°F) controls front-end volatility. Not less than 10%, but not more than 40% of the fuel must evaporate at that temperature. The minimum value ensures that volatility is adequate for normal cold starting. The maximum value controls vapor lock, fuel system vent losses, and carburetor icing.

2. The requirement that at least 50% of the fuel be evaporated at 105°C (221°F) ensures that the fuel has even distillation properties and does not consist of low-boiling and high-boiling components only. This provides control over the rate of engine warm-up and stabilization of slow-running conditions.

3. The requirement that the sum of the 10 and 50 percent evaporated temperatures exceed 135°C (307°F) also controls the overall volatility and indirectly places a lower limit on the 50 percent point. The clause is an additional safeguard against excessive fuel volatility.

4. The requirement that a minimum of 90% of the fuel be evaporated at 135°C (275°F) controls the proportion of less volatile fuel components and, therefore, the amount of unvaporized fuel passing through the engine manifold into the cylinders. The limit represents a compromise between ideal fuel distribution characteristics and commercial considerations of fuel availability which could be affected adversely by further restriction of this limit.

5. The final distillation temperature of 170°C (338°F) maximum excludes any undesirable heavy material which could cause fuel maldistribution and also dilution of the crankcase oil.

All spark-ignition engine fuels have a vapor pressure which is a measure of the tendency of the more volatile fuel components to escape from the fuel tank in the form of vapor. When an aircraft climbs rapidly to a high altitude, the atmospheric pressure over the fuel is reduced and may become less than the vapor pressure of the fuel at its prevailing temperature. If this occurs, the fuel will "boil," and considerable quantities of the more volatile components will escape as vapor through the tank vents.

Vapor pressure for aviation gasolines is controlled and determined by the ASTM Test for Vapor Pressure of Petroleum Products (Reid Method) (D 323/IP 69). Limits are between 38 and 49kPa (5.5 to 7.0 psi). The lower limit is an additional check on adequate volatility for engine starting. The upper limit controls excessive vapor formation during high-altitude flight and "weathering" losses in storage.

Density and Heat of Combustion

No great variation in either density or heat of combustion occurs in modern aviation gasolines since they depend on hydrocarbon composition which is already closely controlled by other specification properties. Both factors have relatively greater importance with jet fuels as discussed in detail later.

Freezing Point

Maximum freezing point values are set for all types of aviation fuel as a guide to the lowest temperature at which the fuel can be used without risk of separation of solidified hydrocarbons. Such separation could lead to fuel starvation through clogging of fuel lines or filters or loss in available fuel load due to retention of solidified fuel in the tanks. The low freezing point requirement also virtually precludes the presence of benzene which, while a high octane material, has a very high freezing point.

The standard freezing-point test involves cooling the fuel until a slurry of crystals form throughout the fuel and noting the temperature at which all crystals disappear on rewarming the fuel. Freezing points are determined by ASTM Test for Freezing Point of Aviation Fuels (D 2386/IP 16).

Storage Stability

Aviation fuel must retain its required properties for long periods of storage in all kinds of climates. Unstable fuels oxidize and form polymeric oxidation products which remain as a resinous solid or "gum" on induction manifolds, carburetors, valves, etc. as the gasoline is evaporated. Formation of this undesirable gum must be limited strictly, and it is assessed by the existent and accelerated (or potential) gum tests.

The existent gum value is the amount of gum actually present in the fuel at the time of the test. It is determined by ASTM Test for Existent Gum in Fuels by Jet Evaporation (D 381/IP 131). The accelerated gum test, ASTM Test for Oxidation Stability of Aviation Fuels (Potential Residue Method) (D 873/IP 138), predicts the possibility of gum forming during protracted storage and decomposition and precipitation of the antiknock additive.

To ensure that the strict limits of the stability specification clauses are met, aviation gasoline components are given special refining treatments to remove the trace impurities responsible for instability. In addition, limited quantities of approved oxidation inhibitors are added. Currently, little trouble is experienced with gum formation or degradation of antiknock additive.

Sulfur Content

Total sulfur content of aviation gasoline is limited to 0.05 percent mass maximum because most sulfur compounds have a deleterious effect on the antiknock efficiency of alkyl lead compounds. If sulfur content were not limited, specified antiknock values would not be reached for highly leaded grades of aviation fuel. Sulfur content is estimated by ASTM Test for Sulfur in Petroleum Products (Lamp Method) (D 1266/IP 107) or X-Ray Spectrographic Method (D 2622).

Some sulfur compounds can have a corroding action on the various metals of the engine system. Effects vary according to the chemical type of sulfur compound present. Fuel corrosivity is assessed by its action on a copper strip used in ASTM Test for Detection of Copper Corrosion from Petroleum Products by the Copper Strip Tarnish Test (D 130/IP 154).

Water Reaction

The original intent of the water reaction test was to prevent the addition of high octane and water soluble components such as alcohol to aviation gasoline. The test methods involved shaking 80 mL of fuel with 20 mL of water under standard conditions and observing phase volume changes and interface condition. Many specifications for aviation gasoline now have phase separation requirements in addition to those for volume change and interface condition. Water Reaction of Aviation Fuels (D 1094/IP 289) rates all three of these criteria.

Automotive (Motor) Gasoline—Use In Aircraft

In general and at the date of this printing, reciprocating aviation engines and the fuel systems in aircraft so powered are designed to operate on one of the grades of fuel specified in ASTM Specification for Aviation Gasolines (D 910), or equivalent. Most major aviation piston engine manufacturers specifically exclude motor gasoline from their list of approved fuels. Many fuel manufacturers also disapprove of the use of motor gasolines in any aircraft. The suitability of motor gasoline for use in aircraft is limited for both technical and safety reasons which are explained below.

Motor gasoline can vary in both composition and quality from supplier to supplier, from country to country, and, in temperate climates, from season to season; in comparison to aviation gasoline, motor gasoline is not a closely or uniformly specified product. A particular variable in recent years is the increasing inclusion of strong detergent additives and of alcohols and/or other oxygenates in motor gasoline.

Availability and cost considerations have encouraged many owners of light aircraft to seek acceptance of motor gasoline as an alternative to aviation gasoline. In recognition of this trend and in order to maintain regulation and control of motor gasoline use, various civil aviation regulatory agencies around the world have extended supplemental or special certification provisions to permit the use of motor gasoline in a limited number of specified aircraft types which are considered, because of design features, to be less sensitive to fuel properties. In the United States of America, such supplemental type certificates (STCs) specify motor gasoline meeting the requirements of ASTM Specification for Automotive Gasoline (D 439). However, the responsibility for any consequences arising from the adoption of alternative fuels such as motor gasoline rests with the owner/operator of the aircraft, the parties who have sought and received approval, and the regulatory agencies that granted said approvals.

The compositional and property differences between motor gasoline and aviation gasoline are detailed below in relation to their potential adverse effects on engine/aircraft operation and flight safety. These factors should be reviewed and evaluated before use of motor gasoline in aircraft.

1. Motor gasolines have a wider distillation range than aviation fuels. This could promote poor distribution of the high antiknock components of the fuel in some carbureted engines. Further, the octane ratings of motor gasoline and aviation gasoline are not comparable due to the different test methods used to rate the two types of fuels. Preignition and detonation conditions could develop due to the appreciable difference in actual antiknock performance of motor and aviation fuels of apparent similar octane ratings.

2. Higher volatility and vapor pressures of motor gasolines could overtax the vapor handling capabilities of certain engine/airframe combinations and could lead to vapor lock or carburetor icing. Fire hazards could also be increased.

3. Motor gasoline has a shorter storage stability lifetime than aviation gasoline and can form gum deposits which can induce poor mixture distribution and other engine mechanical side effects such as valve sticking.

4. Due to higher aromatics content and the possible presence of oxygenates, motor gasoline could have solvent characteristics which are unsuitable for some aircraft engine/airframe combinations. Seals, gaskets, flexible fuel lines, and some fuel tank materials could be affected.

5. Motor gasoline may contain additives which could prove incompatible with certain in-service engine or airframe components. The concentration of additives such as detergents is being continually revised to meet the requirements of advanced automotive fuel injection systems. Alcohols or other oxygenates could increase the tendency for the fuel to hold water, either in solution or in suspension. Other additives, not considered here, could also lead to problems not specifically addressed in this document.

6. The testing and quality protection measures applied to automotive gasoline are much less stringent than for aviation fuels. There is a greater possibility of contamination occurring and less possibility of it being detected. Because motor gasolines meet less stringent requirements, compositional extremes still meeting D 439 might cause undefined difficulties in certain aircraft. Furthermore, D 439 is being continually revised.

7. The antiknock compounds used in leaded motor gasolines contain an excess of chlorine and bromine-containing lead scavengers, whereas aviation gasolines contain a lesser concentration of bromine compounds only. Chlorine compounds give more corrosive combustion products. In addition, lead phasedown regulations in some countries may result in motor gasoline containing insufficient lead to prevent excessive valve seat wear in certain engines.

The above factors illustrate that use of motor gasoline in aircraft may involve certain risks that the potential user must assess.

AVIATION TURBINE FUELS (JET FUELS)

Background

Aircraft gas-turbine engines require a fuel with quite different properties from those for aviation gasoline. Probably the greatest difference is that antiknock value is of no importance and is replaced by the need for a heating fuel of good combustion characteristics and high-energy content. Illuminating kerosine was chosen as the fuel for the first generation of engines largely because of its ready availability, low-fire hazard, good combustion properties, and, not least, the war-time need to conserve gasoline supplies. As engine and fuel system designs have become more complicated, so have the fuel specifications become more varied and restrictive.

Jet fuel quality worldwide is dictated on the commercial side largely by the British Ministry of Defence (DERD) specifications and those of the airlines, engine manufacturers, and industry groups such as ASTM and the International Air Transport Associations (IATA). At airports around the world, jet fuel for airlines is delivered frequently from jointly operated systems in which fuel from a number of suppliers is comingled. This practice has led to the development of a Joint Fueling Systems Check List, which embraces the most critical requirements of the major specifications.

Military jet fuel is dictated largely by the U.S. Department of Defense (U.S. MIL) specifications and corresponding DERD specifications. Grades of commercial and military fuels are virtually identical in basic properties and differ mainly in the types of additives permitted. The only significant exception is in the case of the fuel types used in the Soviet Union and most East European countries. These grades are based on USSR state standards (GOST specifications) and differ in several major respects from their nearest "Western" equivalents.

In the People's Republic of China, early grades of aviation turbine fuel were also based on USSR Standards, but, for recently introduced grades, Western standards and test methods are being adopted.

Only two basic types of jet fuel are in general use worldwide: the kerosine type and the wide-cut gasoline type. The former is a modified development of the illuminating kerosine originally used in gas-turbine engines. The latter is a wider boiling-range material which includes some gasoline fractions, developed in the United States of America primarily for military use, to improve on availability from crude oil. In addition, a number of specialized fuel grades are required for limited military use either as referee fuels or, more particularly, in special high-performance military aircraft.

Composition and Manufacture

Aviation turbine fuels are manufactured predominantly from straight-run kerosines, or kerosine/naphtha blends in the case of wide-cut fuel, from the atmospheric distillation of crude oil. Straight run kerosine from some sweet crudes will meet all the requirements of the jet fuel specification without further refinery processing, but for the majority of crudes, the kerosine fraction will contain trace constituents which have to be removed before the kerosine is merchantable as jet fuel. This is normally effected by hydrotreating (hydrofining) or by a chemical sweetening process (for example, Merox). For further detail on composition and processes to remove undesirable constituents, refer to the section on specification requirements.

Traditionally, jet fuels have been manufactured only from straight run components since the incorporation of raw, thermally or catalytically cracked stocks would invariably produce an off-specification fuel. In recent years, however, hydrocracking processes have been introduced which produce high-quality kerosine fractions ideal for jet fuel blending.

Fuel Developments

Jet fuel development has differed somewhat in Europe and America, but, in recent years, the British and U.S. specification requirements have been brought gradually into line in the interests of military (North Atlantic Treaty Organization—NATO) standardization. Because of the differences in early development philosophy, a brief historical review of the way the quality requirements have developed in different countries is a necessary preamble to a discussion of the test requirements and their significance.

British Jet Fuels

The British jet fuel specification DERD 2482, issued shortly after World War II, was based on the experience gained from operations with illuminating kerosine. It was rather restrictive on aromatics (12 percent max), sulfur content (0.1 percent max), and calorific value (18 500 Btu/lb min) but contained no burning quality requirements. Although further experience permitted relaxation of some of the early requirements, it became necessary to introduce additional tests as new service problems were encountered and amend some existing test limits. For example, the progressive development of more powerful turbine engined aircraft with greater range and altitude performance made the −40°C (−40°F) freezing point limit of the DERD 2482 type fuel inadequate to ensure that solidified hydrocarbons would not separate from fuel during prolonged cold soaking at altitude. A new DERD 2494 specification was therefore issued in 1957 with a maximum freezing point of −50°C (−58°F). This fuel quality remained the optimum compromise between engine development requirements, fuel cost, and strategic availability until very recently. A flash point of 38°C (100°F) minimum was specified, more for fiscal than technical reasons.

While DERD 2494 (Avtur) is now the standard British civil jet fuel, a new DERD 2453 specification (Avtur/FSII) was issued in 1967 incorporating anti-icing [fuel system icing inhibitor (FSII)] and corrosion inhibitor additives in line with the latest U.S. military and NATO requirements. During 1980 a freezing point relaxation to −47°C was permitted in these two specifications.

A less volatile kerosine fuel for use in naval carrierborne aircraft with a flashpoint of 60°C (140°F) min was defined originally by the DERD 2488 specification. In answer to the need for improved low-temperature performance, a later specification, DERD 2498 (Avcat), introduced a modified freezing-point requirement of −48°C (−55°F) max, as compared with the 40°C (−40°F) for the former DERD 2488 specification. In 1966 the freezing point was changed to −46°C (−51°F) maximum, and in 1976 DERD 2452 (Avcat/FSII) was issued to bring the British military high-flash naval fuel into line with U.S. military and NATO requirements.

Because crude oils giving high-gasoline yields are not in abundant supply, wide-range turbine fuel was never used in the United Kingdom to the extent it is in the United States. However, the DERD 2486 (Avtag) specification was introduced in 1951 to cover a fuel grade basically identical to the American wide-cut JP-4 grade (MIL-T-5624). This fuel grade has been used widely, particularly under NATO arrangements. Later, the grade was brought completely into line with the American equivalent with the issue in 1967 of the DERD 2454 specification (Avtag/FSII) which incorporates anti-icing and corrosion inhibitor additives.

Table 4 lists the British DERD specifications for aviation fuels and related products.

American Military Jet Fuels

In the United States, jet fuel development followed a different pattern. The early U.S. specification for JP-1 (MIL-F-5616) called

TABLE 4. British military specifications for aviation fuels and related products.[a]

DERD Specification	British Joint Services Designation	NATO Code Number	Type	Use
2451	AL-31/AL-41	S-748/S-1745	EGME/Di-EGME	fuel system icing inhibitor (FSII)
2452	Avcat/FSII	F-44	high-flash kerosine with FSII	naval carrier aircraft
2453	Avtur/FSII	F-34	kerosine with FSII	standard military fuel
2454	Avtag/FSII	F-40	wide-cut with FSII	standard military fuel
2461	. . .	. . .	various commercial additives	fuel soluble corrosion inhibitor/lubricity additive
2482	Avtur 40	F-30	kerosine, −40°C freeze point	obsolete
2485	Avgas	. . .[a]	aviation, gasoline (4 grades)	civil and military
2486	Avtag	F-45	wide-cut fuel	military
2488	Avcat 40	F-42	high-flash kerosine −40°C freeze point	obsolete
2491	WTA, AL-28	S-1744	water, methanol/water mixtures	thrust augmentation fluid
2492	Avpin	S-746	isopropyl nitrate	turbine engine starter fuel
2494	Avtur	F-35	kerosine, −47°C freeze point	civil
2498	Avcat	F-43	high-flash kerosine, −46°C freeze point	naval carrier aircraft

[a]See Table 1.

for a paraffinic kerosine with a freeze point of −60°C (76°F). These restrictive requirements limited the availability of the fuel, and the grade soon became obsolete (although the term JP-1 is still widely used, quite incorrectly, to describe any kerosine-type fuel). JP-1 has been superseded by a series of military wide-cut fuels.

The wide-range distillate type of turbine fuel originated in the United States where it was realized that in times of military emergency the fuel supply could be increased considerably if more of the readily available gasoline fractions were incorporated into jet fuel to supplement the basic kerosine component. This philosophy is the converse of the World War II choice of kerosine for the early jet engines which was chosen in order to conserve gasoline stocks for the then predominant piston-type aircraft engines.

The first wide-cut grade (JP-2) allowed a Reid Vapor Pressure of 2 psi max to increase availability through the inclusion of heavy gasoline fractions. Experience suggested that a further increase in volatility might be tolerated to give an advantage in engine starting, and the specification of JP-3, therefore, was introduced with vapor pressure limits of 5.0 to 7.0 psi. However, operational problems due to high-venting losses at altitude led finally to the formulation of the JP-4 (MIL-T-5624) specification in 1950 with vapor pressure limits of 2.0 to 3.0 psi (14 to 21 kPa). With slight modifications to the specific gravity and distillation requirements, plus the inclusion of certain additives, JP-4 remains the main fuel for military jet aircraft in the United States and many other countries.

Kerosine type jet fuels have not achieved any significant military use in the United States with the exception of JP-5 (covered also by the MIL-T-5624 specification), a low-volatility kerosine fuel used by naval aircraft during shipboard operations. JP-6 (MIL-F-25656) was a light kerosine type fuel of improved thermal stability intended for a supersonic bomber but has been declared obsolete. JP-7 (MIL-T-38219) is another low-volatility, high-flash point kerosine with enhanced combustion and thermal stability properties that was developed as a fuel for very high-performance (Mach 3) aircraft.

Increasing concern over combat hazards when using JP-4 wide-cut fuel led the Department of Defense to reconsider the use of standard kerosine type fuel. As a result, a new specification for grade JP-8 (MIL-T-83133) was published in draft form in 1972. Service trials were successfully completed, but it was not until 1976 that the specification was officially issued; at the same time the quality requirements were revised in several ways to bring them much closer into line with the long-established British DERD 2494/2453 grades and the civil Jet A-1 in use internationally (it being recognized that, under emergency conditions, practical considerations would necessitate military use of the widely available standard kerosine type fuel). The U.S. Air Force in the United Kingdom has used JP-8 since 1980, and, after many years protracted debate, all the NATO forces in Europe have agreed to switch over to the kerosine-type fuel (NATO grade F-34); the changeover should be complete in 1987.

This change on the European continent was effected primarily for NATO standardization purposes, but elsewhere the advantages gained from switching from JP-4 to JP-8 may be outweighed by the undesirable impact on world jet fuel supplies that might occur if a major international user should change from a wide-cut fuel to a kerosine-type fuel. The issue is further complicated by a general relaxation of the commercial Jet A-1 fuel's freezing point (see below) to a maximum level unacceptable to the U.S. Air Force for certain critical missions. Table 5 lists the U.S. military specifications for jet fuels and related products.

American Civil Jet Fuels

The basic civil jet fuel specification used in the United States is ASTM Specification for Aviation Turbine Fuels (D 1655) which outlines requirements for three types of fuel: Jet A, a nominal −40°C freeze-point kerosine; Jet A-1, a nominal −47°C freeze-point kerosine; and Jet B, a wide-cut gasoline grade (similar to JP-4 but without certain additives). Detailed requirements for aviation turbine fuels as defined in D 1655 appear in Table 6. (Footnotes indicate to what extent Jet A-1 as supplied internationally differs from D 1655.)

Jet A with its −40°C freeze point limit is available only in the United States and accounts for about half the civil jet fuel used throughout the world. It satisfies the requirements of both domestic flights and the majority of international flights originating in the United States. Jet A-1 was originally specified at −50°C freeze point to satisfy the unusual demands of long-range high-altitude flights; in 1980 the freeze point limit was changed to −47°C to respond to availability concerns and to better definitions of long-range flight require-

TABLE 5. U.S. Military jet fuel and related specifications.

Specification	First Issued	Grade	Type	Use
AN-F-32	1944	JP-1	very low-freeze kerosine	obsolete
MIL-F-5616	1950			
AN-F-34	1945	JP-2	wide cut (RVP max 2 psi)	obsolete
AN-F-58	1947	JP-3	wide cut (RVP 3 to 7 psi)	obsolete
MIL-F-5624	1950			
MIL-T-5624	1950	JP-4	wide cut (RVP 2 to 3 psi)	air force standard
MIL-T-5624	1950	JP-5	high-flash kerosine	naval carrier aircraft
MIL-F-25524		JP-TS	thermally stable kerosine	flight test fuel
MIL-F-25656	1956	JP-6	light kerosine (thermally stable)	supersonic bombers (obsolete)
MIL-T-38219	1965	JP-7	low-volatility kerosine (special properties)	very high performance aircraft
MIL-T-83133	1976	JP-8	kerosine (Jet A-1 type)	air force standard
			RELATED SPECIFICATIONS	
MIL-F-5161		...	referee JP-4 and JP-5	ground test fuels (obsolete)
MIL-F-5572		Avgas	aviation gasoline (several grades)	military standard (see Table 1)
MIL-I-25017		...	chemical materials	fuel soluble corrosion inhibitor
MIL-F-25558		RJ-1	high-density kerosine	air force ramjet fuel
MIL-P-25576		RP-1	narrow-cut kerosine	rocket fuel
MIL-I-27686		FSII	ethylene glycol monomethyl ether	fuel system icing inhibitor
MIL-F-81912			high flash, narrow-cut kerosine	missle fuel (navy)
MIL-F-82522		RJ-4	T-H, dimethyl cyclopentadienes	ramjet fuel (navy)
		RJ-5	T-H, Norbornadiene dimer	missile fuel
		RJ-6	63% RJ-5, 37% JP-10	missile fuel
MIL-I-85470		FSII	di-EGME	fuel system icing inhibitor
MIL-P-87107		JP-9	blend of MCH, JP-10, and RJ-5	missile fuel
MIL-P-87107		JP-10	T-H, dicyclopentadiene	propellant/fuel component

TABLE 6. Soviet jet fuel specifications.

Specification	Grade	Type	Use
GOST 10227	T-1	kerosine from low sulfur crude (SR)[a]	general
GOST 10227	TS-1	kerosine from high sulfur crude (SR)[a]	
GOST 10227	TS-1 (premium)	kerosine from high sulfur crude (HT)[b]	most common civil
GOST 10227	T-2	wide-cut fuel	military (?)
GOST 16564	RT	kerosine (HT)[b]	civil
GOST 12308[c]	Thermostable	heavy kerosine, low aromatics, low sulfur (HT)[b]	special military (?)
GOST 9145	T-5, T-6	obsolete	

NOTE: TS-1 and RT appear as TC-1 and PT, respectively, in the Russian alphabet.
[a]SR = straight run.
[b]HT = hydrotreated.
[c]Earlier issues of GOST 12308 covered grades T-6 and T-7, now presumed obsolete.

ments. For international flights outside the United States, Jet A-1 is the standard grade. To differentiate from the military fuel grades (which often contain special additives not used commonly in civil fuels), the terms Jet A-1 and Jet B are now used worldwide to describe the basic kerosine and wide-cut gasoline types of civil jet fuel although the latter has very limited civil usage at present.

A number of jet fuel specifications also are issued by the major U.S. aircraft engine manufacturers and certain airlines. These are either similar to the equivalent ASTM grades or are less restrictive versions of one or more ASTM grades.

Soviet Jet Fuels

A wide range of jet fuels covered by various Soviet GOST specifications are manufactured for both civil and military use. The main grades are covered also by similar specifications issued by a number of other East European countries. While Soviet jet fuel characteristics in some cases differ considerably from those of jet fuels used elsewhere, the main properties are controlled by test methods very similar to their ASTM/IP equivalents. A few additional test methods, for example, iodine number (related to the olefin content), hydrogen sulfide content, ash content, and naphthenic acid soaps are sometimes included. Thermal stability is usually specified also by a completely different test procedure.

Only limited information is available on some of these fuel grades and only a few are in regular civil use. Brief details are shown in Table 6. TS-1 and more recently RT (plus their East European equivalents) are the only grades normally offered to international airlines at civil airports. While the quality of TS-1 has often been found inferior in several respects to Western Jet A-1 type fuel, the latest RT grade, now becoming available at major civil airports in the Soviet Union, will probably satisfy current jet A-1 specification requirements except for flash point (28°C specified minimum).

Other National Jet Fuels

Several other countries also issue jet fuel specifications and the most important of these are listed in Table 7. In most cases these specifications are identical with their American or British equivalents, particularly for countries committed to multinational military standardization agreements (for example, NATO). In most of these countries, little or no practical use is made of the national specifications (which often tend to be out of date against international aviation standards) and most jet fuel is normally manufactured as Jet A-1 to the commercially accepted international "Check List" standard. However, a few countries (for example, Australia, Brazil, France, and Sweden) do make considerable use of their own national standards, and the Canadian jet fuel specifications differ in several respects from the "Check List" and in fact include one quite different grade (Jet A-2 with a minimum flash point of 33°C).

International Standard Specifications

Modern civil aviation recognizes few frontiers and there is, therefore, a need for aviation fuels having similar characteristics to be available in all parts of the world. This is especially important in the case of

TABLE 7. Other national aviation fuel specifications.

Country/Grade	Kerosine	Wide Cut	High-Flash Kerosine
Australia/DEF (Aust)	5208	. . .	207
Brazil/CNP	QAV-1	QAV-4	. . .
Canada/CAN2-	3.23	3.22	3.24
China (Peoples Republic)	GB1788	SY1009	SY1010
France/AIR-	3405	3407	3404
West Germany/TL-9130-	. . .	0006	. . .
Italy/AER-M-C	141	142	143
Japan/JIS-K-	2209	2209	. . .
Sweden/FSD-	8607	8608	. . .

the jet fuels used by international airlines, and, to provide a suitable basis, "guidance material" has been prepared by the International Air Transport Association (IATA) and issued in the form of two specifications, for a kerosine and a wide-cut gasoline-type jet fuel. The fundamental requirements of these two IATA grades are identical to those covered by the main American and British specifications, but a number of options are left to user choice, particularly in respect of additive requirements, and the specifications are not quite so restrictive in some respects. Many major airlines have now adopted the IATA Guidance Material and some issue it under their company designation.

Despite efforts to standardize jet fuels, there are still a number of differences in minimum quality standards even among the major internationally used specifications. These differences are of little significance to military organizations because they normally purchase jet fuel against their own national specifications. Similarly, domestic airlines are not affected since their fuel supplies are relatively uniform within the country in which they fly. International air carriers do experience problems because they often purchase more fuel offshore than they use domestically.

When purchasing jet fuel in other countries, international airlines are faced with a confusing array of names, definitions, and specifications for jet fuels. As a partial solution, airlines quote their Jet A-1 requirements against either the DERD 2494, ASTM D 1655, or IATA specifications. Jet B requirements are stated against DERD 2486, U.S. Mil-T-5624, ASTM D 1655, or IATA specifications. Even with these specifications as guidelines, allowable additive contents frequently are defined poorly.

For the above reasons, a need arose for basic fuel specifications which define the quality requirements precisely and in such a manner that the most stringent requirements of each of the major international specifications are included. This ensures that aviation fuels meeting these specified requirements will automatically be in full compliance with any or all of the commonly quoted official specifications relating to that particular grade and will therefore be acceptable to the widest possible range of users. Suitable specifications have been drawn up by a working party representative of all major international oil companies supplying aviation fuels outside of North America and have been issued as the "Aviation Fuel Quality Requirements for Jointly Operated Systems" (AFQRJOS). This document consists of two Joint Fueling Systems "Check Lists" giving the requirements for kerosine type Jet A-1, and two grades of aviation gasoline.

Specification Requirements

The requirements for jet fuels stress a different combination of properties and tests than those required for aviation gasoline. The same basic controls are needed for such properties as storage stability and corrosivity, but the gasoline antiknock tests are replaced by tests directly and indirectly controlling energy content and combustion characteristics. Therefore, it is convenient to deal with the chemical properties (and composition) separately from the physical properties of jet fuel.

Chemical Properties and Composition

Jet fuels consist entirely of hydrocarbons except for trace quantities of sulfur compounds and approved additives. Since jet fuels are produced normally by blending straight-run distillate components, they contain virtually no olefins. Olefins are limited by specifications and usually determined by ASTM Test for Hydrocarbon Types in Liquid Petroleum Products by Fluorescent Indicator Adsorption (D 1319/IP 156) although olefins in some specifications are controlled by the ASTM Test for Bromine Number of Petroleum Distillates and Commercial Aliphatic Olefins by Electrometric Titration (D 1159/IP 130).

The amounts of aromatics are also limited directly because they are not so clean burning as other hydrocarbon types. High aromatics can cause smokiness and carbon deposition in the engines and increase the luminosity of the combustion flame, which can adversely affect the life of certain designs of combustion chamber. During the crude oil shortages of the 1970s, jet fuel availability was jeopardized in many regions unless aromatics content was permitted to rise above 20% vol. Specifications were modified to allow suppliers to release fuel batches up to 22% aromatics for Jet A-1 or 25% for Jet A provided they rendered periodic reports of such batches to the airlines to permit monitoring of engine conditions. Higher aromatics fuel released in this way is often referred to as "reportable" fuel.

As well as having adverse effects on combustion performance, very high aromatics content (above ca. 30%) can cause deterioration of aircraft fuel system elastomers, which may result in fuel leakage. Aromatics content also is determined by Method D 1319/IP 156.

A reasonably good correlation has been found between total aromatic content and fuel hydrogen content as determined by Hydrogen Content of Aviation Turbine Fuels by Low Resolution Nuclear Magnetic Resonance Spectrometry (D 3701/IP 338). As a result mass percent hydrogen is being considered as an alternative to aromatic content in Specification D 1655.

The principal nonhydrocarbon components permitted in jet fuels are sulfur compounds. The amount and type of sulfur compound varies with crude source, but normally there is no difficulty in meeting the specified total sulfur content, which ranges in level between 0.2 and 0.4 percent by mass maximum. ASTM Methods D 1266/IP 107, D 1552, or D 2622 are used for total sulfur.

Experimental evidence indicates high-sulfur levels may adversely affect the carbon-forming tendency in combustion chambers, and the presence of large amounts of sulfur oxides in the combustion gases may lead to corrosion problems. Direct corrosivity of sulfur compounds in the liquid phase is controlled by the copper strip test (ASTM D 130/IP 154), although this particular procedure is not always capable of reflecting fuel corrosivity towards all metals used in fuel systems. For example, service experience with corrosion of silver components in certain engine fuel systems led to the development of a Silver Corrosion Test (IP 227), which now appears in the British military specifications for kerosine type turbine fuels.

Mercaptan sulfur content of jet fuels is limited to a maximum in the range of 0.001 to 0.005 percent by mass because of objectionable odor, adverse effect on certain fuel system elastomers, and corrosiveness toward fuel system metals. Mercaptan sulfur content is determined by ASTM Test for Mercaptan Sulfur in Gasoline, Kerosine, Aviation Turbine, and Distillate Fuels (Potentiometric Method) (D 3227/IP 342) or by the go/no go Doctor Test (D 235/IP 30).

Mercaptan removal from fuel blends can be effected by sweetening processes which convert them to less objectionable, involatile disulfides. Most sweetening processes use a chemical reagent which does not affect any reduction in total sulfur content. The more common hydrofining procedure, however, catalytically converts all sulfur compounds to gaseous hydrogen sulfide which is then stripped from the refined fuel. Improper control of any sweetening process can produce trace materials which will subsequently cause service trouble even though the fuel meets all the specified physical and chemical tests.

No direct limit is placed on the presence of oxygenated materials, but if they are present as acidic compounds (such as phenols and naphthenic acids) they are controlled in different specifications by a variety of acidity tests. The total acidity by ASTM Test for Neutralization Number by Color-Indicator Titration (D 974/IP 139) is still permitted in some specifications but has been found to be insufficiently sensitive to detect trace acidic materials which can, among other things, adversely effect the water separating properties of fuel. The D 974 method has, therefore, now been modified to make it more sensitive; the new method is designated D 3242/IP 354 (ASTM Test Method for Acidity in Aviation Turbine Fuel) and is of similar sensitivity to the earlier acidity test method (IP 273) formerly used in British military specifications. The acidity limits specified normally vary with the test method required.

Oxygen-containing impurities in the form of gum are limited by the ASTM Test for Existent Gum in Fuels by Jet Evaporation (D 381/IP 131) although in practice currently refined straight-run jet fuels have gum values very much lower than the specification limits. Because of this, some specification authorities have been considering deletion of the test as unnecessary, but it has so far been retained because of its value as a quality control check during distribution (the test providing an easy indication of any contamination by heavier fuels or lubricants during shipment). Due to the inherently good oxidation stability of jet fuels, it is, however, no longer normal practice to include the potential gum test (ASTM D 873/IP 138) in jet fuel specifications.

The various approved additives for jet fuels include oxidation inhibitors to improve storage stability, metal deactivators to neutralize the adverse effect of copper on fuel stability, and corrosion inhibitors intended for the protection of storage tanks and pipelines. An anti-icing additive (fuel system icing inhibitor) is called for in many military fuels, and an electrical conductivity improver (antistatic additive) may be required to minimize fire and explosion risks due to electrostatic discharges in installations and equipment during pumping operations. Details of the various approved additives (mandatory or optional) are included in every specification and discussed in later paragraphs.

Volatility

Volatility is one of the more obvious differences between kerosine and wide-cut jet fuels. The low-volatility requirements for kerosine are controlled by flash point and a small number of distillation points determined by ASTM Test for Distillation of Petroleum Products (D 86/IP 123). Distillation tests and the ASTM Test for Vapor Pressure of Petroleum Products (Reid Method) (D 323/IP 123) control the higher volatility of wide-cut jet fuels. The overall effect of the volatility tests is to give a reasonable compromise between safety, combustion efficiency, and fuel availability.

Distillation points of 10, 20, 50, and 90 percent are specified in various ways to ensure that a properly balanced fuel is produced with no undue proportion of light or heavy fractions. The distillation end point excludes any heavy material which would give poor fuel vaporization and ultimately affect engine combustion performance. For some American military fuels the standard "ASTM Distillation" (D 86) can be replaced by a gas chromatography method (ASTM D 2887), but different test limits are then specified.

The vapor pressure control (on wide-cut fuels) is related to engine starting (both cold on the ground and in-flight relighting at altitude) and also limits fuel tank venting and possible vapor lock at altitude.

Flash point is a guide to the fire hazard associated with the fuel. Flash point can be determined by several test methods which are not always directly comparable. The minimum flash point in British military specifications is usually defined by the Abel Method (IP 170), except for high-flash kerosine where the ASTM Test for Flash Point by Pensky-Martens Closed Tester (D 93/IP 34) is specified. The ASTM specification for aviation turbine fuels calls for Test for Flash Point by Tag Closed Tester (D 56), or ASTM Test for Flash Point by Setaflash Closed Tester (D 3828/IP 303).

It should be noted that the various flash point methods can yield different numerical results, and in the case of the two most commonly used methods (Abel and TAG) it has been found that the former (IP 170) can give results up to 1 to 2°C lower than the latter method (ASTM D 56). Setaflash (D 3828/IP 303) results are generally very close to Abel values.

Density and Heat of Combustion

The density of a fuel is a measure of the mass per unit volume. It is used in fuel load calculations since weight or volume fuel limitations (or both) may be necessary according to the type of aircraft and flight pattern involved. In most cases, the volume of fuel which can be carried is limited by tank capacity. To achieve maximum range, a high-density fuel is preferred because it provides the greatest heating value per unit volume of fuel. The heating value per unit mass of fuel, however, falls slightly with increasing density so that on occasions where the weight of fuel which can be carried is limited (for example, to achieve maximum pay load), it can be advantageous to use a lower density fuel provided adequate tank volume capacity is available.

Density and heat of combustion (calorific value) vary somewhat according to crude source. Paraffinic fuels have a slightly lower density but higher gravimetric colorific value than those from naphthenic crudes. Naphthenic fuels have superior calorific values on a volume basis (joules/litre or Btu/gal).

As fuel density varies with temperature it must be specified under standard conditions and the most usual are 15°C or 60°F. Density at 15°C (kg/m^3) is now becoming the most widely used unit for fuel density worldwide although some specifications still employ relative density (or specific gravity) 15.6°/15.6°C (60°/60°F). Relative density is the ratio of the mass of a given volume of fuel to the same volume of water under specified conditions. ASTM Test for Density, Relative Density (Specific Gravity), or API Gravity of Crude Petroleum and Liquid Petroleum Products by Hydrometer Method (D 1298/IP 160) may be used to determine density and relative density.

In the United States, it is common to specify fuel density in terms of American Petroleum Institute (API) gravity. API gravity may be determined directly by use of ASTM Test for API Gravity of Crude Petroleum and Petroleum Products (Hydrometer Method) (D 287) or calculated from relative density (specific gravity) by the following formula

$$\text{API gravity, degrees} = \frac{141.5}{\text{sp gr } 15.6/15.6°\text{C } (60/60°\text{F})} - 131.5$$

No reference temperature is needed since 15.6°C (60°F) is specified in the formula.

Heat of combustion is the quantity of heat liberated by the combustion of a unit quantity of fuel with oxygen. Heat of combustion affects the economics of engine performance. The specified minimum value is normally a compromise between the conflicting requirements of maximum fuel availability and good fuel consumption characteristics. ASTM Test for Heat of Combustion of Liquid Hydrocarbon Fuels by Bomb Calorimeter (D 240/IP 12) or ASTM Test for Heat of Combustion of Hydrocarbon Fuels by Bomb Calorimeter (High-Precision Method) (D 2382) provide a means for direct measurement of fuel energy content.

An alternative criterion for energy content is the "aniline gravity product" (AGP) which is related to calorific value and more easily determined. AGP is the product of the gravity (expressed in degrees API) and the aniline point of the fuel as determined by ASTM Test for Aniline Point and Mixed Aniline Point of Petroleum Products and Hydrocarbon Solvents (D 611/IP 2). Aniline point is the lowest temperature at which the fuel is miscible with an equal volume of aniline and is inversely proportional to the aromatic content. The relationship between AGP and calorific value is given in ASTM Test for Estimation of Net Heat of Combustion of Aviation Fuels (D 1405/IP 193 or D 4529).

In another empirical method (D 3338) the heat of combustion is calculated from the fuel's density, the 10%, 50%, and 90% distillation temperatures, and the aromatic content. This method avoids the use of aniline, a highly toxic reagent. Neither calculation method is acceptable in case of disputes which can only be resolved by conducting D 2382 for fuels meeting D 1655 or D 240 for fuels meeting the British specifications.

Low-Temperature Properties

Jet fuels must have acceptable freezing points and low-temperature pumpability characteristics so that adequate fuel flow to the engine is maintained during long cruise periods at high altitude. ASTM Test for Freezing-Point of Aviation Fuels (D 2386/IP 16) and its associated specification limits guard against the possibility of solidified hydrocarbons separating from chilled fuel and blocking fuel lines, filters, nozzles, etc.

The viscosity of fuels at low temperature is limited to ensure that adequate fuel flow and pressure are maintained under all operating conditions and that fuel injection nozzles and system controls will operate down to design temperature conditions. Viscosity can affect significantly the lubricating property of the fuel which has an influence on fuel pump service life. Viscosity is determined by ASTM Determination of Kinematic Viscosity of Transparent and Opaque Liquids (and the Calculation of Dynamic Viscosity) (D 445/IP 71).

Combustion Quality

Combustion quality is largely a function of the hydrocarbon composition of the fuel. Paraffins have excellent burning properties in contrast to those of aromatics (particularly the heavy polynuclear types). Naphthenes have intermediate combustion characteristics that are nearer to those of paraffins. Because of compositional differences, jet fuels of the same class can vary widely in burning quality as measured by smoke formation, carbon deposition, and flame radiation.

Smoke point is determined by ASTM Test for Smoke Point of Aviation Turbine Fuels (D 1322). The smoke-point test alone is not universally accepted as a reliable criterion for combustion performance. An alternative test is the ASTM Test for Luminometer Numbers of Aviation Turbine Fuels (D 1740). This test was developed because combustion chamber life is shortened in certain jet engines if a fuel produces luminous flames that result in high metal temperatures. The luminometer number test apparatus is essentially a smoke-point lamp modified to include a photoelectric cell for flame radiation measurement and a thermocouple arrangement to measure temperature rise across the flame.

Another alternative to the smoke-point alone is a combination of smoke-point and fuel naphthalenes content as determined by Test Method for Naphthalene Hydrocarbons in Aviation Turbine Fuels by Ultraviolet Spectrophotometry (D 1840). In this test naphthalenes content is obtained by measuring absorption in the ultraviolet range of a solution of the fuel at known concentration. A fourth alternative, used in some specifications, is hydrogen content (D 3701/IP 338).

However, the relationships of all these tests to the engine performance parameters listed above are completely empirical and do not apply equally to different engine designs, particularly where major differences in engine operating conditions exist.

Thermal Stability

Although the conventional (storage) stability of aviation fuel has long been defined and controlled by the existent and accelerated gum tests, another test is required to measure the stability of a fuel to the thermal stresses to which it is subjected in an aircraft fuel system. During flight, fuel receives considerable heat input because it is employed as a cooling medium for engine oil, hydraulic and air conditioning equipment, etc. Consequently fuels must not form lacquers or deposits which can adversely affect the efficiency of oil/fuel heat exchanges, metering devices, fuel filters, and injector nozzles. This property becomes increasingly important as more fuel-efficient aircraft engines are developed, where operating temperatures are higher while fuel flows are lower. In supersonic aircraft, there is additional heat input from kinetic heating of the airframe.

Early research on the problem indicated the need for a dynamic test and led to the development of the ASTM-CRC Fuel Coker (ASTM D 1660) which is a laboratory test apparatus for assessing the tendency of jet fuels to deposit thermal decomposition products in fuel systems. Although the fuel coker test provides valuable data, it requires a large fuel sample and has relatively poor precision. More recently, ASTM Test for Thermal Oxidation Stability of Aviation Turbine Fuels (JFTOT Procedure) (D 3241/IP 323) has been developed that partially overcomes these disadvantages. The JFTOT (Jet Fuel Thermal Oxidation Tester) has now superseded the Fuel Coker in many fuel specifications and is listed as an alternative method to the Fuel Coker in the ASTM jet fuel specification.

Contaminants

Modern aircraft fuel systems demand a fuel free from water, dirt, and foreign contaminants. To deliver contaminant-free fuel a number of multistage filtration systems are employed at terminals, airports, and as part of fuel service vehicles. Jet fuel is widely distributed from refineries to airports through pipelines that also handle other products, particularly in the United States. As a consequence, contamination of jet fuel by water, solids, and additive traces is inevitable and must be removed by ground filtration systems. Foreign materials are apt to be surface active and by dispersing water and dirt interfere with proper operation of filtration systems. To remove surfactants it is common practice to provide clay filtration at pipeline terminals and airports.

Testing for contaminants of various types occurs at many points in the distribution system. During aircraft fueling itself, jet fuel is tested for "clear and bright" by examining visually a sample using Method D 4176. Delivered fuel must also be below 1 mg/L particulate and 30 mg/L free water according to IATA and U.S. military specifications.

ASTM Test for Particulate Contaminant in Aviation Turbine Fuels (D 2276/IP 216) provides field quality control of dirt content. It can be supplemented by a visual assessment of membrane appearance after test against ASTM color standards which are described in Appendix A3 of D 2276/IP 216 and in Practice for Filter Membrane Color Ratings of Aviation Turbine Fuels (D 3830). However, no direct relationship exists between particulate content weight and membrane color, and field experience is required to assess the results by either method.

Free water dispersed in jet fuels can be detected with a variety of field test kits developed over the years by major oil companies. These tests generally rely on color changes produced when chemicals go into aqueous solution. ASTM Test for Undissolved Water in Aviation Turbine Fuels (D 3240) has been standardized and employs a device called the Aqua-Glo II which is capable of more precise quantitative results than chemical tests although simplicity is sacrificed.

The total water content of aviation fuels (free and dissolved water) can be determined by the ASTM Test for Water in Liquid Petroleum Products by Karl Fischer Reagent (D 1744). However, the basic procedure has inadequate precision, and improvements are needed.

Water, Retention and Separating Properties

Because of their higher density and viscosity, jet fuels tend to retain fine particulate matter and water droplets in suspension for a much longer time than aviation gasoline.

Jet fuels also can vary considerably in their tendency to pick up and retain water droplets or to hold fine water hazes in suspension depending on the presence of trace surface active impurities (surfactants). Some of these materials (such as sulphonic and naphthenic acids and their sodium salts) may originate from the crude source or from certain refinery treating processes. Others may be picked up by contact with other products during transportation to the airfield, particularly in multiproduct pipelines. These latter materials may be natural contaminants from other less highly refined products (for example, heating oils) or may consist of additives from motor gasolines (such as glycol-type anti-icing agents). It should be noted that some of the additives specified for jet fuel use (for example, corrosion inhibitors and static dissipator additives) may also have surface active properties.

TABLE 8. Detailed requirements for aviation turbine kerosine.[a]

COMPOSITION		
Total acidity, mg KOH/g	max	0.1[b]
Aromatics, % vol	max	20.0[c]
Sulfur, total, % mass	max	0.30
Sulfur, mercaptan, % mass	max	0.003
or Doctor Test		negative

TABLE 8. Detailed requirements for aviation turbine kerosine[a] *(continued).*

VOLATILITY		
Distillation		
Initial boiling point °C (°F)		Report
Fuel recovered		
10% vol at °C (°F)	max	204 (400)
20% vol at °C (°F)		report
50% vol at °C (°F)		report
90% vol at °C (°F)		report
End point, °C (°F)	max	300 (572)
Residue, % vol	max	1.5
Loss, % vol	max	1.5
Flash point, °C (°F)	min	38 (100)
Relative density (sp gr) 15/15°C (60°F/60°F)	min	0.755
	max	0.840
or Gravity, °API	max	51
	min	37
FLUIDITY		
Freezing point, °C	max	−40 Jet A
		−47 Jet A-1
Viscosity at −20°C, cSt (mm²/s)	max	8.0
COMBUSTION		
Calorific value, net, btu/lb (MJ/kg)	min	18 400 (42.8)
Smoke point, mm	min	25
or luminometer number	min	45
or smoke point, mm	min	20[d]
and naphthalenes, % vol	max	3.0
CORROSION		
Corrosion, copper (2 h at 100°C)	max	1
STABILITY		
Thermal stability (JFTOT)		
Filter pressure differential, mm Hg	max	25.0
Tube deposit rating (visual)		less than Code 3
or Thermal stability (Fuel Coker)[e]		
Filter pressure drop ins Hg	max	3.0
Preheater Deposit		less than Code 3
CONTAMINANTS		
Existent gum, mg/100 mL	max	7
Water reaction		
Interface rating	max	1b
Separation rating	max	2
CONDUCTIVITY		
Electrical conductivity, pS/m[f]	min	50
	max	450
PERMITTED ADDITIVES		
Antioxidant, mg/L (lb/1000 bbl)[g]	max	24 (8.4)
Metal deactivator, mg/L (lb/1000 bbl)	max	5.7 (2.0)
Static dissipator additive, mg/L[h]		
ASA-3	max	1.0
or STADIS 450	max	3.0

[a]ASTM Specification for Aviation Turbine Fuels D 1655.
[b]In international specifications, maximum is 0.015 mg KOH/g.
[c]Or 25 per cent vol max "reportable."
[d]Or 18 mm min "reportable."
[e]This method permitted only in ASTM D 1655; not permitted in other major fuel specifications.
[f]Applies only if fuel is doped with static dissipator additive.
[g]In international specifications, a minimum of 17 mg/L antioxidant is mandatory in hydrotreated fuels.
[h]Mandatory in international specifications.

The presence of surfactants can also impair the performance of the water separating equipment (filter/separators) used in fuel handling systems to remove traces of free (undissolved) water. Very small traces of free water can affect adversely jet engine and aircraft operations in several ways, for example by ice formation. The water retention and separating properties of jet fuels have become a critical quality consideration in recent years.

The standard water reaction test (ASTM D 1094/IP 289) for jet fuels is the same as for aviation gasolines, but the interface and separation ratings are defined more critically. Test assessment is by subjective visual observation. Although sufficiently precise when made by an experienced operator, the test can cause rating difficulties under borderline conditions. As a consequence, a more objective and sensitive test now is included in many specifications; this is ASTM Test for Water Separation Characteristics of Aviation Turbine Fuels (D 2550). A later, portable version of this procedure is the ASTM Test for Determining Water Separation Characteristics of Aviation Turbine Fuels by Portable Separometer in the Field (D 3948). In both of these tests a controlled fuel-water emulsion is pumped through miniature filter-coalescer pads, and the amount of water remaining in the filtered fuel is measured optically.

Field tests for surfactants such as the portable separometer, sometimes referred to as the microsep, are often used to monitor the effectiveness of clay filtration systems installed to remove surfactants from fuel during distribution. Water separation tests (D 2550 or D 3948) are also included in some fuel specifications as a means of ensuring that fuel, as manufactured, is free from surfactants and has inherently good water separation characteristics.

Another contamination problem is that of microbiological growth activity which can give rise to service troubles of various types. This problem generally can be avoided by the adoption of good housekeeping techniques, but major incidents in recent years have led to the development of several microbiological monitoring tests for aviation fuel. In one of these, fuel is filtered through a sterile membrane which is cultured subsequently for microbiological growths. Other tests employ various techniques to detect the presence of viable microbiological matter, but none of the tests have yet been standardized.

Electrical Conductivity

Many fuel specifications require the use of static dissipator additive (see below) to improve safety in fuel handling. In such cases the specification defines both minimum and maximum electrical conductivity; the minimum level ensures adequate charge relaxation while the maximum prevents too high a conductivity, since this can upset some capacitance-type fuel gauges in aircraft.

The standard field test for electrical conductivity has been the ASTM Test for Electrical Conductivity of Aviation and Distillate Fuels Containing a Static Dissipator Additive (D 2624/IP 274). Although the method was intended for the measurement of the conductivity of fuel at rest in storage tanks, it can also be used in a laboratory on fuel samples. However, a more precise laboratory method is ASTM Method for Electrical Conductivity of Liquid Hydrocarbons by Precision Meter (D 4308).

Miscellaneous Properties

A few specifications for some of the lesser used fuel grades call for special tests not generally applied to aviation fuels. Examples of these seldom required tests are: the explosiveness test for JP-5 fuel (Federal Test Method 1151) and color limits as determined by the Saybolt Method (ASTM D 156). Although not normally a specification item, color deterioration can sometimes be a useful indication of interproduct contamination or instability (gum formation).

Inspection Data on Aviation Turbine Fuels

Many airlines, government agencies, and petroleum companies make detailed studies of inspection data provided on production aviation turbine fuels. Because a large number of inspections generally are involved, these studies are made frequently with the aid of a computer. Without a standardized format for reporting data from different sources, transcribing the reported data for computer programming is laborious.

To facilitate the reporting of inspection data on aviation turbine fuels, a standardized report form has been established by ASTM. It appears as Appendix X.2 to ASTM Specification for Aviation Turbine Fuels (D 1655) and is shown as Appendix I to this chapter. This format has found broad acceptance by suppliers of both U.S. military and civil fuels.

AVIATION FUEL ADDITIVES

General

Only a limited number of additives are permitted in aviation fuels, and for each fuel grade the type and concentration are controlled closely by the appropriate fuel specifications. Additives may be included for a variety of reasons, but, in every case, the specifications define the requirements as follows:

Mandatory—Must be present between given minimum and maximum limits.

Permitted—May be added by fuel manufacturer's choice up to a maximum limit.

Optional—May be added only with agreement of user/purchaser, within specified limits.

Not allowed—Additives not listed in specifications are not permitted.

In the case of aviation gasolines, there is little significant variation in the types and concentrations of additives normally present in each standard grade, but considerable variations occur in the additive content of jet fuels depending on whether they are for civil or military use and on the country of origin. Table 9 summarizes the most usual additive content of aviation fuels on a worldwide basis (except for Soviet grades). Many exceptions occur and reference to the appropriate specifications should be made to establish the precise requirements.

Additive Types

Additives may be included in aviation fuels for a number of reasons; while their purpose is generally to improve certain aspects of fuel performance, they usually achieve the desired effect by preventing or suppressing some undesirable fuel behavior, such as corrosion, icing, oxidation, detonation, etc. The effectiveness of additives is due to their chemical nature and the resulting interactions with constituents of the fuel (usually trace constituents). It is important when approving additives not only to establish that they achieve the desired results and are fully compatible with all materials likely to be contacted, but to ensure also that they do not react in other ways to produce adverse side effects on fuel performance in an engine or on its behavior in aircraft or ground handling systems (possibly by interfering with the action of other additives present). The basic testing of aviation additives for approval purposes is usually carried out by

TABLE 9. Summary of usual additive requirements for British and U.S. aviation fuels.

Additive	Aviation Gasoline	Civil Jet Fuels	Military Jet Fuels
Color dyes	mandatory	no	No
Tetraethyl lead	permitted	no	No
Antioxidant	mandatory	permitted[a]	permitted[a]
Metal deactivator	no	permitted	permitted
Corrosion inhibitor	normally no[b]	no[c]	mandatory
Fuel system icing inhibitor	normally no[d]	no[c]	mandatory
Static dissipator additive	normally no[e]	permitted[f]	. . .[g]
Lubricity additive	no	no[c]	mandatory[h]

[a]Mandatory for fuels produced by hydrogen treating processes in British and major U.S. military grades and in international civil Jet A-1.

[b]Not allowed by British gasoline specification. Permitted at user option in U.S. military gasoline specification, not by civil gasoline specifications.

[c]By special customer request only.

[d]User option in ASTM D 910, but, if required, would normally be added by aircraft operator.

[e]Permitted only in Canadian and British aviation gasolines.

[f]Mandatory in Canada and in international civil Jet A-1; permitted in ASTM D 1655 specification.

[g]Mandatory in British Military jet fuels DERD 2494/2453 and in U.S. grades JP-4 and JP-8.

[h]Normally, lubricity additives and corrosion inhibitors are the same additive.

individual aircraft and engine manufacturers; their results and conclusions may then appear in company fuel specifications, but they are also reviewed for adoption by the international fuel specification authorities. In the case of civil aviation fuels (ASTM and IATA) additive approval policy then usually involves wide consultation followed by a consensus decision, but for military fuels (US-MIL and DERD specifications) additive approvals may be quite arbitrary and are frequently designed to satisfy particular military considerations, often related to individual equipment types or operational needs.

As a valuable step towards rationalizing the approval procedure for aviation fuel additives, ASTM Practice for Evaluating the Compatibility of Additives with Aviation-Turbine Fuels and Aircraft Fuel System Materials (D 4054) has been drawn up. Used in conjunction with ASTM "Guidelines for Additive Approval" (Research Report D02-1125) and "Compatibility Testing with Fuel System Materials" (Research Report D2-1137), the procedure is intended primarily for approval of additives covered by the ASTM jet fuel specification (D 1655) but is likely to be used widely in the future by other bodies (for example, equipment manufacturers and national agencies) as a basis for official approvals.

The following paragraphs describe the aviation fuel additives in current use. No attempt is made to fully list the various chemical and trade names of the approved materials since these are detailed in the appropriate specifications.

Tetraethyl Lead (TEL)

Tetraethyl lead is used widely to improve the antiknock characteristics of gasoline. An adverse side effect of this additive is the deposition in the engine of solid lead compounds formed in the combustion process which promote spark plug fouling and corrosion of cylinders, valves, etc. To alleviate this potential problem, a scavenging chemical—ethylene dibromide—is mixed with the TEL. Ethylene dibromide largely converts the lead oxides into volatile lead bromide which is expelled with the exhaust gases. As a compromise between economic considerations and the avoidance of side effects, the maximum addition of TEL is controlled carefully by specifications using ASTM tests for lead in gasoline (D 2547, D 2599, or D 3341). TEL is not permitted in jet fuels, as lead compounds present even in trace amounts could damage turbine blades and other hot engine components.

Color Dyes

Dyes are used to identify the different grades of aviation gasoline. The required colors are achieved by the addition of various combinations of up to three special anthraquinone-based and azo dyes (blue, yellow, and red). The amounts permitted are controlled between closely specified limits to obtain the desired fuel colors. ASTM Test for Color of Dyed Aviation Gasolines (D 2392) is used to determine color. Dyes are not permitted in jet fuels.

Antioxidant (Gum Inhibitor)

Antioxidant additive is mandatory in aviation gasolines to prevent the formation of gum and precipitation of lead compounds. The additive concentration is controlled closely by specifications.

Jet fuels are inherently more stable than gasolines. The use of antioxidants is permitted but not mandatory in all cases. However, it was established some years ago that some jet fuels which are hydrogen treated (or produced by a "hydrogen process") can generate a high-peroxide content which can cause rapid deterioration of nitrile rubber fuel system components and/or the formation of insoluble gums. This is due to the removal of most of the naturally occurring inhibiting materials (trace sulfur compounds) by the hydrogen-desulfurizing treatment. To combat this problem, some specifications make the use of antioxidant mandatory in the case of hydrogen-treated jet fuels. A maximum permitted limit of 24.0 mg/L usually applies for all jet fuels, with a minimum of 8.6 mg/L when the additive is made mandatory.

A wide range of antioxidants are approved, although the chemical types permitted vary under different specifications. These additives are generally phenolic or nitrogen-containing materials.

Metal Deactivator

The use of one approved metal deactivator (N, N′-disalicylidene 1,2-propane diamine) is permitted in jet fuels but not in aviation gasoline. The purpose of this additive is to passivate certain metallic materials in solution in jet fuels, some of which degrade by catalytic action the storage stability (gum-forming tendency) or thermal stability of the fuel. Copper is the worst of these materials and metal deactivator is normally only used in fuels on which a copper sweetening process has been used to remove mercaptans, etc. in order to convert any trace metal carry over to inert copper chelates. In British specifications the copper content (IP 225) of such fuels is strictly limited.

Corrosion Inhibitors

Corrosion inhibitors are used to minimize corrosive rusting of steel pipelines, tanks, etc. in contact with fuels in which traces of water are sometimes present. A direct benefit from the use of corrosion inhibitors is a reduction in the amount of scale and fine rust shed into the fuel as particulate contaminant.

The use of corrosion inhibitors also provides significant improvements in the lubricating properties ("lubricity") of jet fuels. This helps overcome problems involving excessive friction such as rapid component wear or seizure sometimes encountered in certain fuel pumps and other fuel system components. These problems (which are not fully understood) are largely of a metallurgical nature and often can be avoided by a change of component materials. However, fuels having poor lubricating properties appear to aggravate the problem. No single reliable test has yet been devised to measure the "lubricity" of jet fuels.

Because of this lack of knowledge, the specification requirements for fuel soluble corrosion inhibitors have varied during recent years. Different specifications call for corrosion inhibitors for different reasons, and fuel users often impose their own special requirements on basic specifications.

As a general rule, U.S. and British military jet fuels require the addition of an approved corrosion inhibitor. Civil jet fuels, wide-cut or kerosine type, normally do not contain corrosion inhibitors.

A number of proprietary fuel soluble corrosion inhibitors are available, mostly of American manufacture. The basic approval list is presented as a Qualified Products List (QPL-25017) of materials satisfying the requirements of the U.S. military specification MIL-I-25017. A similar list (QPL2461) accompanies the British military specification DERD 2461. These documents include for each approved additive the minimum and relative effective concentrations together with the maximum allowable concentration (different for each additive type).

Most additives can exhibit some undesirable side effects under certain circumstances. Included are adverse effects on water-separating properties, electrical conductivity, and thermal stability. For this reason, there is a reluctance to permit the use of additives that have not been thoroughly proved in service.

Fuel System Icing Inhibitor

A fuel system icing inhibitor (FSII) (anti-icing additive) was used originally to overcome fuel system icing problems encountered by U.S. military aircraft. Unlike most commercial aircraft and many British military aircraft, which have their main fuel filters heated to prevent blockage by ice formed from water precipitated from fuels in flight, U.S. Air Force (USAF) aircraft have no such protection, and, prior to 1960, icing of filters and other fuel system components caused a number of accidents. FSII prevents these problems by lowering the freezing point of any water present in the fuel to such a degree that no ice formation can occur. The additive has only a limited solubility in fuel and a much greater affinity for water.

FSII is now a mandatory requirement in most military fuel specifications, especially those covered by NATO or SEATO standardization agreements. It consists of a pure material "ethylene glycol monomethyl ether" (EGME) which is known also as methyl cellosolve, methyl oxitol, methyl glycol, and 2-methoxyethanol by various chemical manufacturers.

When FSII has been added to supplies of jet fuel for naval carrier aircraft (JP-5/Avcat) it has sometimes been found difficult to meet the minimum fuel flashpoint requirement (60°C/140°F) due to the low flashpoint of EGME (about 40°C). Consequently, a new type of high-flash FSII has now been introduced, consisting of di-EGME with a flash point of about 65°C. This type of FSII, now approved for use in naval and some other grades of military jet fuel, is covered by a new U.S. specification MIL-I-85470 and included with standard EGME in the latest issue of the British DERD 2451 specification.

Shortly after introducing the widespread use of FSII to combat icing problems, the USAF experienced a great reduction in the number of microbiological contamination problems being encountered in aircraft tanks and ground handling facilities. Studies subsequently confirmed that this improvement was due primarily to the biocidal nature of the additive. It is now generally accepted that EGME is a very effective biostat if used continuously in jet fuels.

Commercial aircraft are, with minor exceptions, provided with fuel heaters and have no requirement for an anti-icing additive, particularly in view of the relatively high cost which would be involved. A few turbine-powered helicopters and corporate aircraft are not provided with fuel heaters, but, in most cases, the operators make their own arrangements to inject EGME into the aircraft fuel supplies when necessary.

Several Soviet manufactured jet liners are not fitted with fuel heaters. Domestically, these Soviet aircraft use fuels containing an anti-icing additive usually based on ethyl cellosolve. Elsewhere the standard methyl cellosolve material generally is accepted.

In tropical areas, operators of some types of civil aircraft occasionally require fuel containing FSII for its biocidal properties. In these cases, local arrangements can be made, prior to fueling the aircraft, to inject the additive at the airfield.

Although primarily a jet fuel additive, EGME is sometimes used as an anti-icing additive in aviation gasolines. However, for general aviation piston-engined aircraft, it is more common to use isopropyl alcohol (IPA). ASTM Standard Specification for Fuel System Icing Inhibitors (D 4171) defines the properties of both these materials.

It has been observed when isopropyl alcohol (IPA) is added to Grade 100 Avgas as a fuel system icing inhibitor that the antiknock rating of the fuel may be significantly reduced. Typical performance number reductions with addition of one volume percent IPA have been about 0.5 PN on the lean rating and 3 to 3.5 PN on the rich rating. Nevertheless, there has not been any field evidence or experience reported to date to suggest these reductions have caused engine distress, that is, knocking. The ASTM Specification for Aviation Gasolines (D 910) contains a cautionary statement and gives further details about this phenomenon. In Grade 80 Avgas, addition of the IPA additive may increase the octane quality.

Static Dissipator Additive (Antistatic Additive)

Static charges built up during movement of fuel can lead to high-energy spark discharges capable of igniting flammable fuel/air mixtures. This is true particularly in the case of modern jet fuels because of their extreme purity (very low-natural conductivity), the high-pumping velocities employed, and the use of microfiltration equipment capable of producing a high rate of electrical charge separation and static buildup in the flowing fuel. Static dissipator additives (SDAs), sometimes referred to as conductivity improver additives, are designed to prevent this hazard by increasing the electrical conductivity of the fuel which, in turn, promotes a rapid relaxation of any static charges.

Almost all jet fuel specifications permit the optional use of SDA by agreement between supplier and user, but an increasing number make its use mandatory. SDA is now mandatory for U.S. military grades JP-4 and JP-8 and for British grades DERD 2494/2453. For international Jet A-1 supplies, the AFQRJOS Check List specification also has a mandatory requirement for SDA; the requirement in the ASTM D 1655 specification remains optional and most Jet A supplied to domestic and international carriers in the United States does not contain static dissipator additive. In Canada, static dissipator additive is mandatory in jet fuels and permitted in aviation gasoline since the hazards of static discharge are particularly severe under very low-ambient temperature conditions.

Two static dissipator additives are currently approved for use in aviation fuels—Shell Anti-Static Additive ASA-3 and Du Pont Stadis 450. The Shell additive comprises a mixture of equal parts of the chromium salt of an alkylated salicylic acid (chromium AC), calcium didecyl, sulphosuccinate (calcium Aerosol), and a vinyl/methacrylate copolymer (Alkadine). The composition of the Du Pont additive is proprietary but is known to contain no metals. These additives are used at extremely low dosage levels—typically less than 1 mg/L for ASA/3 or 3 mg/L for Stadis 450—to provide electrical conductivity within the ranges quoted in fuel specifications.

Nonspecification Additives

No additives except those mentioned in the preceding paragraphs are approved under current fuel specifications, but there are others that are sometimes used for special purposes. Only one of these (Biobor JF biocide additive) has any significant use in commercially operated aircraft but several merit attention.

Biocides

Biobor JF is a mixture of dioxaborinanes used to prevent microbiological growth in hydrocarbon fuels. Extensive use in railway diesel fuels established its effectiveness and lack of troublesome side effects. After further satisfactory use by several major airlines, most engine and airframe manufacturers gave limited approval of Biobor JF use on an intermittent or noncontinuous basis in concentrations not to exceed 270 mg/L (20-ppm elemental Boron).

Biobor JF normally is used to "disinfect" aircraft during a period when the aircraft is left standing filled, or partially filled, with doped fuel. The fuel is then used in the normal manner although it may be drained off or diluted with undoped fuel before being burned in the engines. To minimize the possible deposition of boron compounds in the engine, this treatment is only permitted at infrequent intervals.

Antismoke Additives

Although the generation of excessive exhaust smoke by jet engines is primarily an engine combustion design problem, some reduction in smoke density in some engines may be possible by the use of additives. Among these are Ethyl C.I. 2 and Lubrizol 565. The former is approved by one major engine manufacturer for postoverhaul test-cell operation of the engines up to 3 h. Neither is approved for flight use because of the additive ash content.

Ignition Control Additive

To minimize the adverse effects of spark plug deposits in gasoline engines, several phosphorus containing additives were developed. Typical of these is tricresyl phosphate (TCP) (also known as tritolyl phosphate–TTP) which modifies lead deposits so they do not cause preignition.

Spark plug fouling was pronounced in certain older types of aircraft piston engines, and TCP/TTP was used to overcome the problem. As these engines were withdrawn from service and as the permitted TEL content of aviation gasoline gradually was reduced, the problem diminished. Now it is doubtful that this additive has any significant use.

Antimisting Additive

Over a number of years research has taken place to develop a fuel additive which would eliminate the fireball which can be formed in an otherwise survivable crash.

While one additive has been tested extensively major development work remains to be done and opinions on the additive's usefulness are divided.

ADDITIVE TESTS

Although the type and amount of each additive permitted in aviation fuels are limited strictly, test methods for checking concentrations are not always specified. Where tests are not called for, a written statement of the additives addition is accepted as evidence of its presence. The following paragraphs recap the tests for the additives discussed previously.

Tetraethyl Lead

In aviation gasolines, the TEL content has such a critical influence on the antiknock properties and deposit forming tendency of the fuel that a test for TEL content is included in all routine laboratory tests. There are three alternative ASTM test methods for lead in gasoline—D 2547, D 2599, and D 3341.

Color

After the specified amounts of color dyes have been added to aviation gasolines, the color usually is checked by visual inspection. However, the Lovibond Method (IP 17), Federal Test Method 103, and ASTM Color of Dyed Aviation Gasolines (D 2392) are called for in various specifications.

Antioxidant, Metal Deactivator, Corrosion Inhibitor

After the required amounts of antioxidant, metal deactivator, or corrosion inhibitor have been added to aviation fuels, checks normally are not run on concentrations, so no test methods are included in specifications for this purpose. However, the refiner is normally required to state on the Certificate of Quality the amount of each additive used. Occasionally, a need arises to determine the amount of corrosion inhibitor remaining in the fuel. Several analytical methods have been developed, but none have been standardized.

Fuel System Icing Inhibitor

FSII used in jet fuels can be lost by evaporation and also lost rapidly into any water which might contact the fuel during transportation. Several methods are available to determine FSII content. Included among these are Icing Inhibitor in Aviation Turbine Fuels (IP 277), which provides a "referee" test (method C) and, for routine purposes, a simpler version (method D). A simple field test for rapid determination of FSII content is under development.

Static Dissipator Additive

Static dissipator additive is used in such small concentrations that it is extremely difficult to detect by any standard analytical procedure. Therefore, static dissipator additive is controlled by measuring the resulting electrical conductivity of the fuel. This is most commonly done in situ using hand-held meters under the procedure covered by the ASTM Method D 2624/IP 274.

Applicable ASTM Specifications

D 910	Specification for Aviation Gasolines
D 1655	Specification for Aviation Turbine Fuels
D 4171	Specification for Fuel System Icing Inhibitors

Applicable ASTM/IP Standards

ASTM	*IP*	*Title*
D 56		Flash Point by Tag Closed Tester
D 86	123	Distillation of Petroleum Products
D 93	34	Flash Point by Pensky-Martens Closed Tester
	170	Flash Point by Abel Apparatus
D 130	154	Detection of Copper Corrosion from Petroleum Products by the Copper Strip Tarnish Test
D 156		Saybolt Color of Petroleum Products (Saybolt Chromometer Method)
	17	Colour by the Lovibond Tintometer
D 235	30	Doctor Test
D 240	12	Heat of Combustion of Liquid Hydrocarbon Fuels by Bomb Calorimeter
D 323	69	Vapor Pressure of Petroleum Products (Reid Method)
D 381	131	Existent Gum in Fuels by Jet Evaporation
D 445	71	Kinematic Viscosity of Transparent and Opaque Liquids (and the Calculation of Dynamic Viscosity)

Applicable ASTM/IP Standards *(continued)*

ASTM	*IP*	*Title*
D 611	2	Aniline Point and Mixed Aniline Point of Petroleum Products and Hydrocarbon Solvents
D 873	138	Oxidation Stability of Aviation Fuels (Potential Residue Method)
D 909	119	Knock Characteristics of Aviation Gasolines by the Supercharge Method
D 974	139	Neutralization Number by Color-Indicator Titration
	225	Copper in Aviation Turbine Fuels and Light Petroleum Distillates
D 1094	289	Water Reaction of Aviation Fuels
D 1159	130	Bromine Number of Petroleum Distillates and Commercial Aliphatic Olefins by Electrometric Titration
D 1266	107	Sulfur in Petroleum Products (Lamp Method)
D 1298	160	Density, Relative Density (Specific Gravity), or API Gravity of Crude Petroleum and Liquid Petroleum Products by Hydrometer Method
D 1319	156	Hydrocarbon Types in Liquid Petroleum Products by Fluorescent Indicator Adsorption
	227	Silver Corrosion by Aviation Turbine Fuels
D 1322	(57)	Smoke Point of Aviation Turbine Fuels
D 1405	193	Estimation of Net Heat of Combustion of Aviation Fuels
D 1552		Sulfur in Petroleum Products (High-Temperature Method)
D 1660		Thermal Stability of Aviation Turbine Fuels
D 1740		Luminometer Number of Aviation Turbine Fuels
D 1744		Water in Liquid Petroleum Products by Karl Fischer Reagent
	277	Icing Inhibitor in Aviation Turbine Fuels
D 1840		Naphthalene Hydrocarbons in Aviation Turbine Fuels by Ultraviolet Spectrophotometry
D 2276	216	Particulate Contaminant in Aviation Turbine Fuels
D 2382		Heat of Combustion of Hydrocarbon Fuels by Bomb Calorimeter (High-Precision Method)
D 2386	16	Freezing Point of Aviation Fuels
D 2392		Color of Dyed Aviation Gasolines
	224	Trace Amounts of Lead in Aviation Turbine Fuels and Light Petroleum Distillates
D 2547	248	Lead in Gasoline, Volumetric Chromate Method
D 2550		Water Separation Characteristics of Aviation Turbine Fuels
D 2551	171	Vapor Pressure of Petroleum Products (Micromethod)
D 2599	228	Lead in Gasoline by X-Ray Spectrometry
D 2622		Sulfur in Petroleum Products (X-Ray Spectrographic Method)
D 2624	274	Electrical Conductivity of Aviation Fuels and Distillate Containing a Static Dissipator Additive

Applicable ASTM/IP Standards *(continued)*

ASTM	*IP*	*Title*
D 2699	237	Knock Characteristics of Motor Fuels by the Research Method
D 2700	236	Knock Characteristics of Motor and Aviation Fuels by the Motor Method
D 2887		Boiling Range Distribution of Petroleum Fractions by Gas Chromatography
D 3227	342	Mercaptan Sulfur in Gasoline, Kerosine, Aviation Turbine, and Distillate Fuels (Potentiometric Method)
D 3240		Undissolved Water in Aviation Turbine Fuels
D 3241	323	Thermal Oxidation Stability of Aviation Turbine Fuels (JFTOT Procedure)
D 3242	354	Total Acidity in Aviation Turbine Fuel
D 3338		Estimation of Heat of Combustion of Aviation Fuels
D 3341	270	Lead in Gasoline—Iodine Monochloride Method
D 3343		Estimation of Hydrogen Content of Aviation Fuels
D 3602		Water Separation Characteristics of Aviation Turbine Fuels (Field Test)
D 3701	338	Hydrogen Content of Aviation Turbine Fuels by Low Resolution Nuclear Magnetic Resonance Spectrometry
D 3703		Peroxide Number of Aviation Turbine Fuels
D 3828	303	Flash Point by Setaflash Closed Tester
D 3830		Filter Membrane Color Ratings of Aviation Turbine Fuels
D 3948		Determining Water Separation Characteristics of Aviation Turbine Fuels by Portable Separometer
D 4052		Density and Relative Density of Liquids by Digital Density Meter
D 4057		Manual Sampling of Petroleum and Petroleum Products
D 4176		Free Water and Particulate Contamination in Distillate Fuels (Clear and Bright Pass/Fail Procedures)
D 4305		Filter Flow of Aviation Fuels at Low Temperatures
D 4306		Sampling Aviation Fuel for Tests Affected by Trace Contamination
D 4308		Electrical Conductivity of Liquid Hydrocarbons by Precision Meter
D 4529		Estimation of Net Heat of Combustion of Aviation Fuels

Bibliography

Friedman, R., "Aviation Turbine Fuel Properties and Their Trends." SAE Paper 810850, Society of Automotive Engineers, Warrendale, PA, July 1981.

Smith, M., *Aviation Fuels,* G. T. Foulis & Co. Ltd., England, 1970.

Goodger, E. and Vere, R., *Aviation Fuels Technology,* Macmillan Publishers Ltd., England, 1985.

Modern Petroleum Technology, Institute of Petroleum, London, 5th ed. 1984.

Ogston, R., "A Short History of Aviation Gasoline Development, 1903–1980." SAE Paper 810848, Society of Automotive Engineers, Warrendale, PA, July 1981.

APPENDIX I

ASTM **D 1655**

INSPECTION DATA ON AVIATION TURBINE FUELS

(See the back of the form itself or Specification D 1655 Appendix A2 for instructions on use of form)

REPORT DATE ____________ QUANTITY U.S. GALLONS: ____________

CONTRACT No. ____________ SAMPLING LOCATION: ____________

ORDER No. ____________ DESTINATION: ____________

DATE SAMPLED ____________ PRODUCT NAME: ____________

SAMPLE No. ____________ COMPLIES WITH SPECIFICATIONS: ____________ REMARKS: ____________

BATCH No. ____________

TANK No. ____________

GRADE ____________

Method		APPEARANCE	Results
10	D 156	Color (Saybolt)	
20	—	Visual (B=Bright & C=Clear)	
		COMPOSITION	
100	D 974/3242	Acidity, Total (mg KOH/g)	
110	D 1319	Aromatics (vol %)	
120	D 1319	Olefins (vol %)	
130	D 1219/1323	Sulfur, Mercaptan (wt %)	
140	D 484	Doctor Test (P=Pos, N=Neg)	
150	D 1266	Sulfur, Total (wt %)	
		VOLATILITY	
200	D 86	Distillation Init. BP (F)	
205	D 86	" 10% Rec (F)	
210	D 86	" 20% Rec (F)	
215	D 86	" 50% Rec (F)	
220	D 86	" 90% Rec (F)	
225	D 86	" 95% Rec (F)	
230	D 86	" Final BP (F)	
235	D 86	Residue (%)	
240	D 86	Loss (%)	
245	D 86	Recovery at 400 F (%)	
250	FTMS 1151	Explosiveness (vol %)	
260	D 56/3243	Flash Point, Tag Closed (F)	
261	D 93	Flash Point, Pensky Martin (F)	
270	D 1298	Gravity, API (60 F)	
280	D 1298	Gravity, Specific (60/60 F)	
290	D 323	Vapor Pressure (lb Reid)	
		FLUIDITY	
300	D 2386	Freezing Point (F)	
310	D 445	Viscosity at -20 C (cSt)	
		COMBUSTION	
400	D 1405	Aniline Gravity Product	
410	D 1405	Net Heat of Comb. (Btu/lb)	
420	D 1740	Luminometer No.	
430	D 1322	Smoke Point	
440	D 1840	Naphthalenes (vol %)	
450	D 1655	Smoke-Volatility Index	

Method		CORROSION	Results
500	D 130	Copper Strip (2 h at 212 F)	
510	IP 227	Silver Strip	
		STABILITY	
600	D 1660	Coker ΔP (in Hg)	
610	D 1660	Coker Tube Color Code	
611	D 3241	JFTOT at 260C ΔP (mm Hg)	
612	D 3241	JFTOT at 260C Tube Color Code	
613	D 3241	JFTOT at 245C ΔP (mm Hg)	
614	D 3241	JFTOT at 245C Tube Color Code	
		CONTAMINANTS	
700	IP 225	Copper Content (μg/kg)	
710	D 381	Existent Gum (mg/100 ml)	
720	D 2276	Particulates (mg/liter)	
721	Mil-T-5624 J APPA	Filtration time, min	
722	"	vacuum, mm Hg	
723	"	volume fuel filtered, ml	
740	D 1094	Water Reaction Interface Rating	
741	D 1094	" " Separation Rating	
750	D 2550	WSIM	

ADDITIVES	Brand	Results
800 Anti-icing (vol %)		
810 Antioxidant (lb/M Bbl)		
820 Corrosion Inhib (lb/M Bbl)		
830 Metal Deactivator (lb/M Bbl)		
840 Antistatic, mg/liter		

Method		OTHER TESTS	Results
900	D 2624/3114	Conductivity (pS/m)	
901	D 2624/3114	" Temperature (F)	

APPROVED BY ____________ ____________

Company Representative — Authorized Government Representative

ASTM American Society for Testing and Materials
1916 Race St., Phila., Pa 19103
ASTM D1655

6

Fuels for Land and Marine Diesel Engines and for Nonaviation Gas Turbines

INTRODUCTION

THE DIESEL ENGINE is now fully established in a variety of applications on land and in marine use. On land, it serves to power trains, buses, trucks, and automobiles; to run construction, petroleum drilling, and other off-road equipment; and to be the prime mover in a wide range of power generation and pumping applications. At sea, it serves both to provide main propulsion power and to run auxiliaries.

Gas turbine engines also serve in a wide range of applications. Over half the larger industrial gas turbines are in electric-generation use. Other uses include gas pipeline transmission, cogeneration systems, and transportation. In the military, gas turbines power a number of combatant ships both as main propulsion units and as the power source for auxiliary uses.

The quality criteria and methods for testing fuels for land and marine diesel engines and for nonaviation gas turbines are sufficiently similar to address in a common chapter. Obviously, certain criteria and tests will apply to one or the other rather than both. For example, the cetane number, which is a critical property for diesel fuels, is of limited significance for gas turbine fuels.

Diesel Engine

The diesel engine is a high-compression, self-ignition engine. Fuel is ignited by the heat of the high compression, and no spark plug is used. The diesel cycle consists of charging the combustion chamber with air, compressing the air, injecting the fuel which ignites spontaneously, expanding the burned gases, and expelling the products of combustion.

Diesel engines may be designed to operate on a four- or a two-stroke cycle. Each type has advantages and disadvantages, so the choice depends upon the application. The four-stroke cycle has better volumetric efficiency, good combustion characteristics, and positive exhaust gas scavenging. The principal advantage of the two-stroke cycle is compactness in relation to power output.

Diesel engines vary greatly in size, power output, and operating speeds. While some small units develop only a few brake horsepower, at the other extreme there are engines having cylinder diameters as large as 1050 mm (41.34 in.) developing several thousand horsepower per cylinder. Designed sizes and horsepower output continue to increase. Operating speeds are almost as diverse as size and power output. They range from below 100 rpm for some larger engines to 4000 rpm and above for those used in automotive and other vehicle prime mover service. The entire range of diesel engines can be divided into three broad classification groups indicated in Table 1.

It should not be surprising that diesel engines also vary extensively in their requirements for fuel. Selection of the proper fuel is not a simple procedure but depends upon many variables. Among the most important considerations are:

1. Engine size and design.
2. Operating speed and load changes.
3. Frequency of speed and load changes.

TABLE 1. Range of diesel engines.

Classification	Speed Range	Conditions	Typical Applications[a]
Low speed	below 300 rpm	sustained heavy load, constant speed	marine main propulsion; electric power generation
Medium speed	300 to 1000 rpm	fairly high load and relatively constant speed	marine auxiliaries: stationary power generators: pumping units
High speed	1000 rpm or above	frequent and wide variation in load and speed	road transport vehicles: diesel locomotives: construction equipment

[a]There are many other typical applications not listed.

4. Maintenance considerations.
5. Atmospheric conditions.
6. Fuel price and availability.

Each of the foregoing factors plays a part in dictating the fuel to be chosen for a diesel engine. The relative influence of each factor is determined by the specific application and installation involved.

Nonaviation Gas Turbine

The simple-cycle, gas-turbine engine operates on the Brayton or Joule cycle consisting of adiabatic compression, constant pressure heating, and adiabatic expansion. By adding a heat exchanger to transfer heat from the turbine exhaust gas to the combustor inlet air, thermal efficiency can be increased.

Air is compressed in the compressor through axial or centrifugal stages or both and directed toward the combustion chamber. Here, part of the air mixes with vaporized or atomized fuel and supports combustion. The remainder of the air passes around the flame, cooling the metal surfaces and combining with those gases which are rapidly expanding from combustion. The resulting gas stream is then expanded through one or more turbine wheels which drive the compressor and provide the output power. In a simple-cycle gas turbine, the gas is then exhausted; in a regenerative engine, the exhaust gas is directed through a heat exchanger to heat the combustor inlet air.

Gas turbine engines cover a range from 50 to nearly 100 000 hp. The selection of a gas-turbine fuel oil for use in a given gas turbine requires consideration of the following factors: availability of the fuel, design of the gas turbine and fuel handling system, and maintenance and operating requirements.

COMBUSTION PROCESS

Diesel Combustion

The fuel used in all diesel engines—precombustion chamber, direct injection, two cycle, four cycle—passes through the following processes:

1. Storage, pumping, and handling.
2. Filtering.
3. Heating (if necessary).
4. Atomization and mixing with air.
5. Combustion.
6. Power extraction.
7. Heat exchange and exhaust.

Additionally, marine diesel fuels containing residual oil components require centrifuging during the fuel handling process.

The fuel properties are significant in all these processes and particularly influence combustion and resultant energy extraction. In any combustion process, there are at least three basic requirements:

1. Formation of a mixture of fuel and air.
2. Ignition of the fuel/air mixture.
3. Completion of combustion of the fuel/air mixture.

In the diesel engine, these requirements are met as indicated diagrammatically in Fig. 1. Figure 2 shows a typical pressure versus crank angle diagram for a diesel engine combustion chamber.

Prior to the injection of the fuel, air alone is compressed and raised to a high

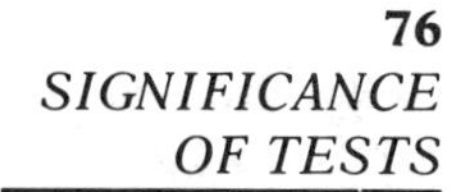

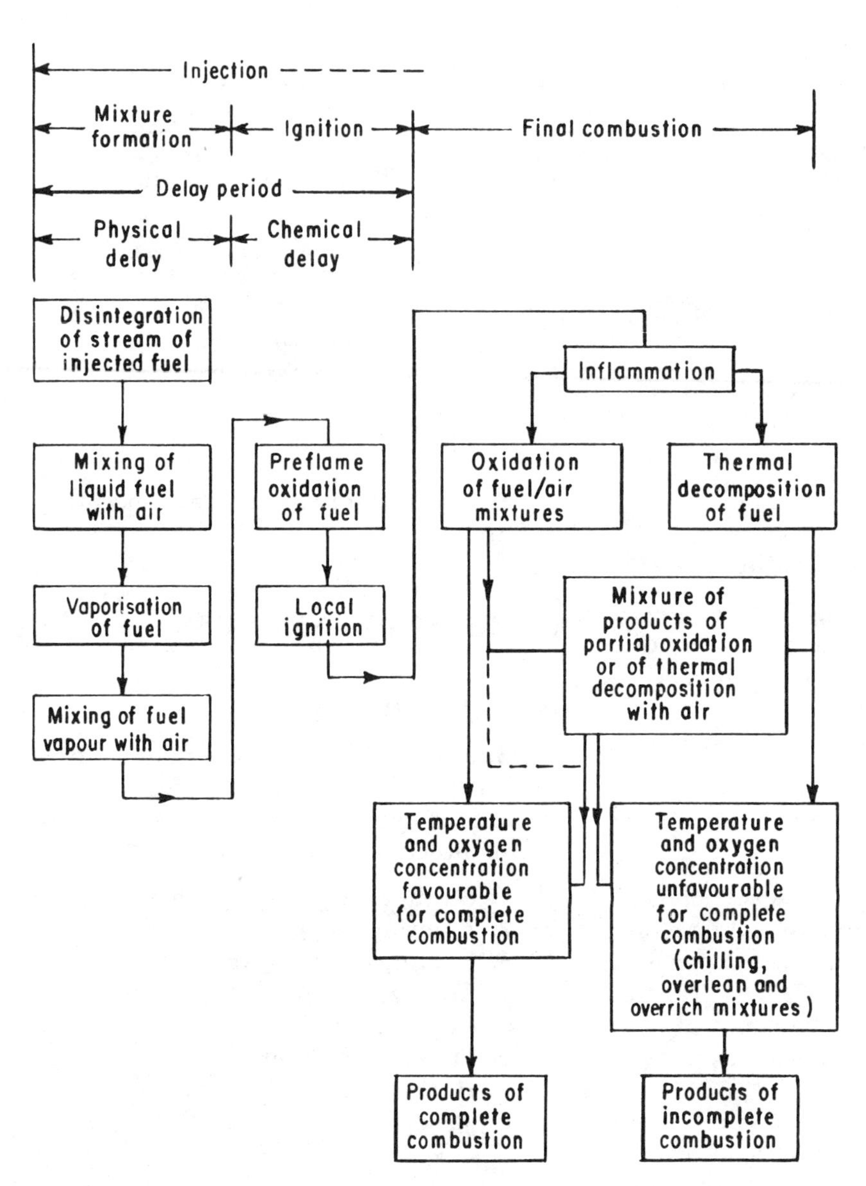

FIG. 1—Outline of the combustion process in the diesel engine. (M. A. Elliot, July 1949). Combustion of diesel fuel, SAE Transactions, Vol. 3, No. 3:492.

temperature during the compression stroke. The final compression pressure and resultant air temperature will vary with compression ratio, speed, and engine design, but a pressure of 31.6 kg/cm^2 (450 psi) and a temperature of 538°C (1000°F) are representative values for naturally aspirated engines. Higher pressures and temperatures occur in blown or supercharged engines. Shortly before the end of compression, at a point controlled by the fuel injection timing system, one or more jets of fuel are introduced into the main combustion chamber or the precombustion chamber.

Ignition does not occur immediately on injection. The fuel droplets must first be vaporized by heat from the compressed air. The fuel/air mixture finally reaches a temperature at which self-ignition occurs, and the flame begins to spread. The duration of the delay period between injection and ignition is controlled by engine design, fuel and air inlet temperatures, degree of atomization of the fuel, and fuel composition. This delay period is known commonly as "ignition delay."

Injection of fuel is continued during the ignition delay. The ignition delay period must be short in order to avoid "diesel knock" which is caused by very rapid burning or detonation of relatively large amounts of fuel gathered in the cylinder before combustion begins. Once the flame has spread completely, the only fuel in the cylinder is that being injected into the burning mixture. This fuel burns almost in-

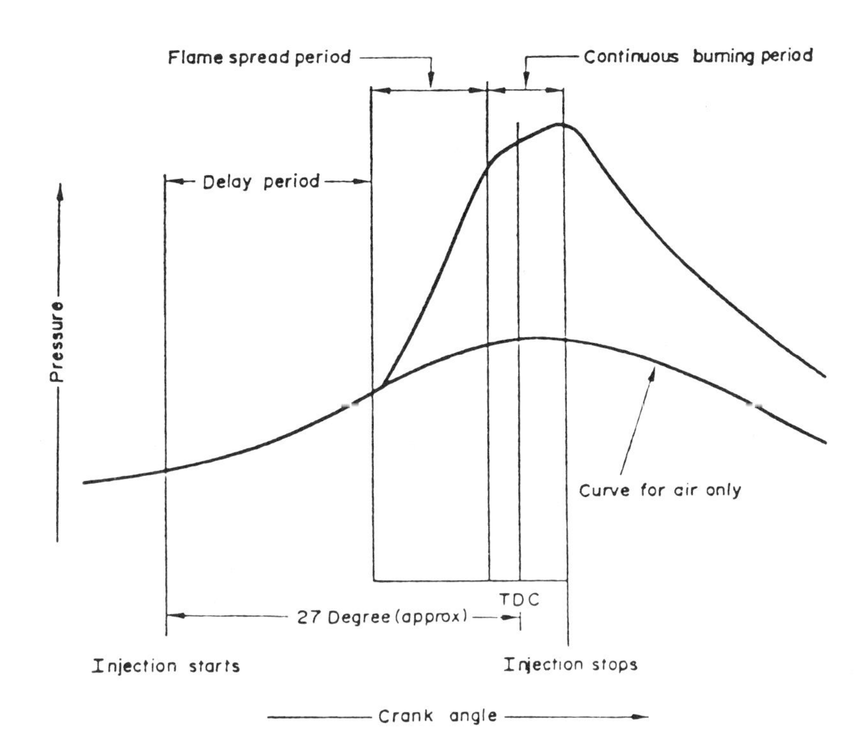

FIG. 2—Pressure variations in a diesel engine cylinder during combustion.

stantaneously. The final part of the combustion cycle is the completion of burning after injection has ceased.

The quantity of fuel, the rate at which it is injected into the engine, and the timing and duration of the injection period all are controlled accurately by a cam-driven injection pump. The pump delivers the fuel to the injectors at a pressure of 130 to 2100 kg/cm^2 (1800 to 30 000 psi) depending upon the design of the injection equipment. Variation in the fuel quantity to conform to different speed or load conditions or both usually is by means of a governor which admits fuel to the combustion chamber at a preset maximum rate until the new conditions are attained. The maximum fuel rate is set to avoid black smoke caused by an excessive amount of fuel.

The maximum amount of air that can be pumped by an engine is determined by its design. The amount of air that can be utilized efficiently determines the optimum injection rate of fuel and, hence, the maximum power output of the engine. Below this maximum, the output of the engine is controlled solely by the amount of fuel supplied.

Pressure charging is used frequently as a means for increasing the amount of air delivered to an engine without increasing its size. A compressor, either directly coupled to the crankshaft (a supercharger) or driven by a turbine using the heat energy in the exhaust gases (a turbocharger), is used to increase the amount of combustion air available. Consequently, the engine is able to burn a greater quantity of fuel. The amount of fuel is ultimately limited by the thermal and mechanical stresses that can be tolerated by engine components.

Nonaviation Gas Turbine Combustion

In a gas turbine engine, fuel is burned continuously at peak cycle pressure to heat the air to moderate temperatures. The combustor is essentially a direct-fired air heater in which fuel is burned with less than one third of the available air, and the combustion products are mixed with the remaining air to cool them to the maximum temperature allowed by metallurgical considerations in the turbine section.

Although gas-turbine combustors vary widely in design arrangement, all perform the same basic functions in much the same way. The two main methods of fuel injection are:

Atomizer

Most combustors employ an atomizer in which fuel is forced under high pressure through an orifice. This breaks up the liquid fuel into small droplets which greatly increases the surface-to-volume ratio. The droplets are then introduced to the primary zone of the combustor where they vaporize and burn; the two processes occur simultaneously.

Prevaporizing (Vaporizing)

This method nearly always relies on direct heating of the fuel. Fuel, together with a proportion of the primary air, a mixture too rich to burn, is supplied to the vaporizing tubes located in the primary zone of the combustor. The fuel is vaporized by the heat released in the primary zone, and the resulting rich mixture is discharged into the primary zone in an upstream direction. The remainder of the primary air is admitted into the primary zone and burning of the mixture occurs.

Combustor inlet air temperature, which depends on engine pressure ratio and load, varies from about 121 to 454°C (250 to 850°F) in various nonregenerative engines. With regeneration, combustor inlet temperatures may be 371 to 649°C (700 to 1200°F). Combustor outlet temperatures range from 649 to 1038°C (1200 to 1900°F) or above. Combustor pressures can range from 3.2 kg/cm^2 (45 psia) to as high as 21.1 kg/cm^2 (300 psia).

GENERAL FUEL CHARACTERISTICS AND SPECIFICATIONS

The basic fuel requirements for land and marine diesel engines and for nonaviation gas turbines are satisfactory ignition and combustion under the conditions existing in the combustion chamber, suitability for handling by the injection equipment, and convenient handling at all stages from the refinery to the engine fuel tank without suffering degradation and without harming any surface which it may normally contact.

Diesel and nonaviation gas-turbine fuels were originally straight-run products obtained from the distillation of crude oil. Today with the various refinery cracking processes, these fuels may contain varying amounts of selected cracked distillates. This permits an increase in the volume of available fuel at minimum cost. The boiling range of distillate fuels is approximately 150 to 400°C (300 to 755°F). The relative merits of the fuel types to be considered will depend upon the refining practices employed, the nature of crude oils from which they are produced, and the additive package (if any) used.

The broad definition of fuels for land and marine diesel engines and for nonaviation gas turbines covers many possible combinations of volatility, ignition quality, viscosity, gravity, stability, and other properties. Various classifications or specifications are used in different countries to characterize these fuels and thereby to establish a framework for definition and reference. ASTM Specification for Diesel Fuel Oils (D 975) is one of the most widely used specifications for diesel fuels. ASTM Specification for Gas Turbine Fuel Oils (D 2880) has some rather similar requirements, as a comparison of Tables 2 and 3, the detailed chemical and physical property requirements for the two classes of fuels, will show.

Blended marine diesel fuels are a relative newcomer to the range of available fuels. The development and use of marine diesel engines for deep-draft vessel propulsion became so pervasive during the 1970s that diesel power generation now predominates in waterborne commerce. Steamships have been displaced from their major role in the world's fleet.

As diesel power came into prominence, equipment manufacturers developed engine designs and fuel handling processes to use the new blended fuels more efficiently. Petroleum refiners, marine fuel suppliers, and vessel operators quickly found that existing specifications and test methods were not completely adequate for their needs. Ad hoc quality and test procedures proliferated. Fuel quality disputes frequently resulted in long, expensive, and inconclusive litigation. The need for a general standard for marine fuels became clear.

To fill the void, the British Standards Institution published BSMA 100, the first generally accepted marine fuel specification, in 1982. This document was superceded in 1987 by the publication of ISO 8217 by the International Standards Organization. ISO 8217 represents a cooperative effort by national standards organizations such as BSI and ASTM to develop an international concensus on minimum marine fuel requirements.

While the development of this concensus was in progress in 1985, ASTM published *Marine Fuels, STP 878.* This STP provides a review of research and developments completed, in progress, or planned up to 1985. The published documents yield an understanding of the problems and complexities of marine fuel production.

ASTM D 975, ASTM D 2880, and ISO 8217 do not provide complete quality specifications for the fuels defined in those documents. In general, they present minimum requirements necessary to assure reasonable operation under the conditions described.

Diesel Fuels

In ASTM D 975, Grades No. 1-D and 2-D are distillate fuels which are the types most commonly used in high-speed engines of the mobile type, medium-speed stationary engines, and railroad engines. Grade 4-D covers the class of more viscous distillates and blends of those distillates with residual fuel oils. No. 4-D fuels are applicable to low- and medium-speed engines employed in services involving sustained load and predominately constant speed.

Examples of equivalent fuel standards, similar to those presented in Table 2, that exist in European countries are DIN 51601 "Fuels for High Speed Diesels" in West Germany and BS 2869 "Petroleum Fuels for Oil Engines and Burners" in the United Kingdom.

TABLE 2. Detailed requirements for diesel fuel oils.[A,H]

Grade of Diesel Fuel Oil		Flash Point, °C (°F)	Cloud Point, °C	Water and Sediment, vol %	Carbon Residue on, 10% Residuum, %	Ash, weight %	Distillation Temperatures, °C (°F) 90% Point		Viscosity Kinematic, cSt[G] at 40°C		Viscosity Saybolt, SUS at 100°F		Sulfur,[D] weight %	Copper Strip Corrosion	Cetane Number[E]
		Min	Max	Max	Max	Max	Min	Max	Min	Max	Min	Max	Max	Max	Max
No. 1-D	A volatile distillate fuel oil for engines in service requiring frequent speed and load changes.	38 (100)	[B]	0.05	0.15	0.01	...	288 (550)	1.3	2.4	...	34.4	0.50	No. 3	40[F]
No. 2-D	A distillate fuel oil of lower volatility for engines in industrial and heavy mobile service.	52 (125)	[B]	0.05	0.35	0.01	282[C] (540)	338 (640)	1.9	4.1	32.6	40.1	0.50	No. 3	40[F]
No. 4-D	A fuel oil for low and medium speed engines.	55 (130)	[B]	0.50	...	0.10	...	...	5.5	24.0	45.0	125.0	2.0	...	30[F]

[A]To meet special operating conditions, modifications of individual limiting requirements may be agreed upon between purchaser, seller, and manufacturer.

[B]It is unrealistic to specify low-temperature properties that will ensure satisfactory operation on a broad basis. Satisfactory operation should be achieved in most cases if the cloud point (or wax appearance point) is specified at 6°C above the tenth percentile minimum ambient temperature for the area in which the fuel will be used. The tenth percentile minimum ambient temperatures for the United States are shown in ASTM D 975 Appendix X2. This guidance is of a general nature; some equipment designs, use flow improver additives, fuel properties, or operations, or a combination thereof, may allow higher or require lower cloud point fuels. Appropriate low temperature operability properties should be agreed upon between the fuel supplier and purchaser for the intended use and expected ambient temperatures.

[C]When cloud point less than −12°C (10°F) is specified, the minimum viscosity shall be 1.7 cSt (or mm^2/s) and the 90 % point shall be waived.

[D]In countries outside the United States, other sulfur limits may apply.

[E]Where cetane number by Method D 613 is not available, Method D 976 may be used as an approximation. Where there is disagreement, Method D 613 shall be the referee method.

[F]Low-atmospheric temperatures as well as engine operation at high altitudes may require use of fuels with higher cetane ratings.

[G]1 cSt = 1 mm^2/s.

[H]The values stated in SI units are to be regarded as the standard. The values in inch-pound units are for information only.

The BS specification covers four classes of fuels marketed specifically as engine fuels. Class A1 is designed primarily as a fuel for higher speed diesel engines; Class A2 is intended as a general purpose diesel fuel; Classes B1 and B2 are for larger engines such as those used in marine practice. Class B2 allows for the inclusion of small amounts of residuum.

While the foregoing classifications deal chiefly with distillate fuels, residual fuels are used extensively in large, slow-speed, marine main propulsion engines and stationary land-based engines principally because of lower fuel cost. More recently, the lower cost aspect has promoted the increasing popularity of residual fuels in some smaller, medium-speed auxiliary engines. Mixtures of residual and distillate fuels are used in installations where a compromise is sought between the more desirable properties of the latter and the lower cost of the former.

ASTM currently has no specification for marine fuels. The international standard is ISO 8217 Marine Fuels—Specifications. This specification has four categories of distillate fuels and fifteen categories of fuels containing residual components.

A typical heavy residual fuel is illustrated in the specifications established by ASTM D 396 to define No. 6 fuel oil, which is often referred to as "Bunker C." These specifications, relatively few in number and broad in their limits, are shown in Table 4. Some diesel engines are operated on fuels of even heavier than No. 6 fuel oil. However, it should be recognized that lower quality fuels, while having the advantage of lower fuel cost, usually require special equipment and special lubricating

TABLE 3. Detailed requirements for gas turbine fuel oils at time and place of custody transfer to user.[F]

Designation[A]	Grade of Gas Turbine Fuel Oil	Flash Point, °C (°F)[B]	Pour Point, °C (°F)[B]	Water and Sediment, vol %	Carbon Residue on 10% Residuum, wt %	Ash, wt %	Distillation Temperature, 90% Point[B] °C (°F)		Kinematic Viscosity, cSt[C] at 40°C (104°F)		Kinematic Viscosity, cSt[C] at 50°C (122°F)	Saybolt Viscosity, S[B] Universal at 38°C (100°F)		Saybolt Viscosity, S[B] Furol at 50°C (122°F)	Specific Gravity 60/60°F (°API)[B]
		Min	Max	Max	Max	Max	Min	Max	Min	Max	Max	Min	Max	Max	Max
No. 0-GT	A naphtha or other low-flash hydrocarbon liquid.	[D]	. . .	0.05	0.15	0.01	. . .	. . .	[D]	. . .	. . .	. . .	. . .	. . .	. . .
No. 1-GT	A distillate for gas turbines requiring a fuel that burns cleaner than No. 2-GT.	38 (100)	−18[E] (0)	0.05	0.15	0.01	. . .	288 (550)	1.3	2.4	. . .	. . .	(34.4)	. . .	0.850 (35 min)
No. 2-GT	A distillate fuel of low ash suitable for gas turbines not requiring No. 1-GT.	38 (100)	−6[E] (20)	0.05	0.35	0.01	282 (540)	338 (640)	1.9	4.1	. . .	(32.6)	(40.2)	. . .	0.876 (30 min)
No. 3-GT	A low-ash fuel that may contain residual components.	55 (130)	. . .	1.0	. . .	0.03	. . .	. . .	5.5	. . .	638	(45)	. . .	(300)	. . .
No. 4-GT	A fuel containing residual components and having higher vanadium content than No. 3-GT.	66 (150)	. . .	1.0	. . .	. . .	. . .	. . .	5.5	. . .	638	(45)	. . .	(300)	. . .

[A]No. 0-GT includes naphtha, Jet B fuel, and other volatile hydrocarbon liquids. No. 1-GT corresponds in general to Specification D 396 Grade No. 1 fuel and Classification D 975 Grade No. 1-D diesel fuel in physical properties. No. 2-GT corresponds in general to Specification D 396 Grade No. 2 fuel and Classification D 975 Grade No. 2-D diesel fuel in physical properties. No. 3-GT and No. 4-GT viscosity range brackets Specification D 396 Grade No. 4, No. 5 (light), No. 5 (heavy), and No. 6 and Classification D 975 Grade No. 4-D diesel fuel in physical properties.

[B]Values in parentheses are for information only and may be approximate.

[C]1 cSt = 1 mm^2/s.

[D]When flash point is below 38°C, or when kinematic viscosity is below 1.3 cSt at 40°C, or when both conditions exist, the turbine manufacturer should be consulted with respect to safe handling and fuel system design.

[E]For cold weather operation, the pour point should be specified 6°C below the ambient temperature at which the turbine is to be operated except where fuel heating facilities are provided. When a pour point less than −18°C is specified for Grade No. 2-GT, the minimum viscosity shall be 1.7 cSt, and the minimum 90% point shall be waived.

[F]Gas turbines with waste heat recovery equipment may require sulfur limits in the fuel to prevent cold-end corrosion (see Appendix X1.4.1.9).

TABLE 4. Requirements for No. 6 fuel oil.[a]

Fuel oil grade	Flash Point, °C (°F)	Water and Sediment, volume %	Saybolt Viscosity, S: Universal 38°C (100°F)		Saybolt Viscosity, S: Furol, 50°C (122°F)		Kinematic Viscosity, cSt, 50°C (122°F)	
	Min	Max	Max	Min	Max	Min	Max	Min
No. 6 Preheating required for burning and handling	60 (140)	2.00[b] ...	(9000) [c]	(900) [c]	300 ...	45 ...	(638) [c]	(92) [c]

[a]These data taken from ASTM Specification D 396 Table 1. *Detailed Requirements for Fuel Oils.*
[b]The amount of water by distillation plus the sediment by extraction shall not exceed 2.00 percent. The amount of sediment by extraction shall not exceed 0.50 percent. A deduction in quantity shall be made for all water and sediment in excess of 1.0 percent.
[c]Viscosity values in parentheses are for information only and not necessarily limiting.

oils to achieve satisfactory performance.

Additives may be used to improve diesel fuel performance. Cetane improvers such as alkyl nitrates and nitrites can improve ignition quality. Pour-point depressants can improve low-temperature performance. Antismoke additives may reduce exhaust smoke which is a growing concern as more and more attention is paid to atmospheric pollution. Antioxidant and sludge dispersants can minimize or prevent the formation of insoluble compounds. Fuels formulated with cracked stock components in particular may form such compounds, which could cause fuel line and filter plugging.

Nonaviation Gas Turbine Fuel

In ASTM D 2880, Grade No. 0-GT includes naphthas, Jet B aviation fuel, and other light hydrocarbon liquids that characteristically have low flash points and low viscosities as compared with kerosine and fuel oils.

In ASTM D 2880, Grade No. 1-GT is a light distillate oil suitable for use in nearly all gas turbines and corresponds in physical properties to No. 1 fuel oil and 1-D diesel fuel oil. Grade No. 2-GT is a heavier distillate grade than No. 1-GT and is for use in turbines not requiring the clean-burning characteristics of No. 1-GT. It is similar in properties to No. 2 fuel oil and 2-D diesel fuel oil.

Grade No. 3-GT is a low-ash fuel having the same viscosity range that bracket No. 4, 5, and 6 fuel oils and 4-D diesel fuel oil. Grade No. 3-GT may be a heavier distillate than that encompassing Grade No. 2-GT, a residual fuel that meets the low-ash requirements, or a blend of a distillate with a residual fuel oil. For gas turbines operating at turbine inlet temperatures below 649°C (1200°F), the nonmandatory vanadium, sodium-plus-potassium, and calcium limits are not critical provided a silicon-based additive or equivalent is used to prevent excessive ash deposition in the turbine. Fuel-heating equipment will be required by the gas turbine in almost every installation using No. 3-GT fuel.

Grade No. 4-GT covers the same viscosity range as Grade No. 3-GT, but it has no restriction on the quantity of ash. Grade No. 4-GT has a suggested magnesium to vanadium weight ratio limit of 3.5 maximum. Nearly all residual fuels will satisfy this requirement, but an additive may be needed to inhibit the corrosive action of the vanadium. In general, Grade No. 4-GT will form ash deposits in the turbine so that, with continuous operation, there will be a progressive reduction in power output and thermal efficiency necessitating periodic shutdown for cleaning.

FUEL PROPERTIES AND TESTS

The properties of a product define its fitness to serve a stated purpose. Once the required properties are determined, they are controlled by appropriate tests and analyses. This section outlines significant fuel quality criteria applicable to diesel engines for land and sea applications, or to nonaviation gas turbines, or to both. Applicable ASTM and Institute of Petroleum (IP) methods for testing and analysis are also given.

Cetane Number

This property is very important in the operation of diesel engines, but it is not important per se in the operation of gas turbine engines except those equipped with a diesel starting engine.

Diesel engine performance is a function of compression ratio, injection timing, the manner in which fuel and air are mixed, and the resulting ignition delay or time from the start of injection to the beginning of combustion. The overall combustion process can be divided into the following stages:

1. Ignition delay.
2. A period of rapid pressure rise.
3. A period of controlled burning.

The rapid pressure rise results from the accumulation of fuel during the ignition delay period and the fact that a large number of ignition points occur throughout the fuel/air mixture. It is this rapid pressure rise that can cause undesirable audible knock, high stresses, and severe engine vibration. Because the rapid pressure rise represents uncontrolled and inefficient combustion, it is desirable to limit ignition delay to a minimum. This limitation can be accomplished mechanically by the selection of a proper nozzle spray pattern, injection pressure, and a combustion chamber designed to create good fuel/air turbulence.

In the majority of diesel engines, the ignition delay period is shorter than the duration of injection. Thus, following the period of rapid pressure rise, the rate of combustion can be controlled to a much greater degree by selection of injection rate, since the fuel is being injected into flame.

The nature of the fuel is also an important factor in reducing the ignition delay. Physical characteristics such as viscosity, gravity, and midboiling point are influential. Hydrocarbon composition is also important as it affects both the physical and combustion characteristics of the fuel. Straight-chain paraffins ignite readily under compression, but branched-chain paraffins and aromatics react more slowly. Since the ignition delay characteristics of diesel fuels directly influence overall engine performance, this property is of primary importance and it is desirable to have a numerical basis for evaluating fuels using this property.

An engine test designed to evaluate fuel ignition delay characteristics was developed in the mid-1930s, the ASTM Test for Ignition Quality of Diesel Fuels by the Cetane Method (D 613). This test involves operating a standard, single cylinder, variable compression ratio engine using a specified fuel flow rate and time of injection (injection advance) for the fuel sample and each of two bracketing reference fuels of known cetane number. The engine compression ratio is adjusted for each fuel to produce a specified ignition delay, and the cetane number is calculated to the nearest tenth by interpolation of the compression ratio values.

To establish the cetane number scale, two primary reference fuels were selected. One, normal cetane, has excellent ignition qualities and, consequently, a very short ignition delay. A cetane number of 100 was arbitrarily assigned to this fuel. The second fuel, alphamethylnaphthalene, has poor ignition qualities and was assigned a cetane number of zero. In 1962, because alphamethylnaphthalene is not stable and a source of constant purity was not available, it was replaced by heptamethylnonane. Heptamethylnonane was calibrated using the two original primary reference fuels and assigned a cetane number of 15. Thus the cetane number scale is now defined by the following equation for volumetric blends of these two primary reference materials

$$\text{Cetane No.} = \text{percent } n\text{-cetane} + 0.15\,(\text{percent heptamethylnonane})$$

In practice, the primary reference fuels are only utilized to calibrate two secondary reference fuels. These are selected diesel fuels of mixed hydrocarbon composition which are designated as "T" and "U". "T" fuel typically has a cetane number of approximately 75 while "U" fuel is usually in the low 20 cetane number range.

Each set of "T" and "U" fuels are paired and test engine calibrations define the cetane numbers for volumetric blends of these two secondaries.

For a given diesel engine, a higher fuel cetane number causes a shorter ignition delay period and a smaller amount of fuel in the combustion chamber when the fuel ignites. Consequently, high cetane number fuels generally cause lower rates of pressure rise and lower peak pressures. Both tend to lessen combustion noise and permit improved control of combustion, resulting in increased engine efficiency and power output.

In addition to these advantages, higher cetane number fuels tend to result in easier starting, particularly in cold weather, and faster warm-up. Reduced exhaust smoke and odor are also associated with higher cetane numbers.

High-speed diesel engines are normally supplied with fuels in the range of 45 to 50 cetane number, although this is trending lower and some requirements may be as low as 35 cetane number. Table 5 summarizes the typical cetane numbers and inspection test characteristics of various marketed diesel fuels grouped on the basis of use. As can be seen, the cetane number ranges from 50 for kerosine to 38 for marine distillate fuel.

Cetane Index

Since the determination of cetane number by engine testing requires special equipment as well as being time consuming and costly, alternate methods have been developed for calculating estimates of cetane number. The calculations are based upon equations involving values of other known characteristics of the fuel.

One of the most widely used methods is based on the Calculated Cetane Index formula. This formula represents a method for estimating the cetane number of distillate fuels from American Petroleum Institute (API) gravity and midboiling point. The index value as computed from the formula is designated as ASTM Calculated Cetane Index of Distillate Fuels (D 976/IP 218). Since the formula is complicated in its manipulation, a nomograph based on the equation has been developed. This nomograph, together with the equation, is shown in Fig. 3. An example illustrating the use of the chart is also presented. It must be recognized that the calculation of cetane index is not an optional method for expressing cetane number. Rather, it is a supplementary tool for predicting cetane number with considerable accuracy when used with due regard for its limitations. The following are among the limitations of calculated cetane index.

1. It is not applicable to fuels containing additives for raising cetane number.

2. It is not applicable to pure hydrocarbons, synthetic fuels, alkylates, or coal tar products.

3. Correlation is fair for a given type of fuel but breaks down if fuels of widely different composition are compared.

4. Appreciable inaccuracy in correlation may occur when used for crude oils,

TABLE 5. Typical inspections of diesel fuels.

Fuel Property	Fuel Type			
	Kerosine	Premium Diesel	Railroad Diesel	Marine Distillate Diesel
Cetane number	50	47	40	38
Boiling range, °C (°F)	163 to 288 (325 to 550)	182 to 357 (360 to 675)	176 to 357 (350 to 675)	176 to 250 (350 to 500) (90 percent)
Viscosity, cSt @ 40°C	33	35	36	47
Gravity, °API	42	37	34	26
Sulfur, weight percent	0.12	0.30	0.50	1.2
Uses	high speed engines for city buses	high speed engines for buses, trucks, tractors, light marine applications	medium speed engines for buses, railroad, marine, and stationary applications	low speed engines for buses, heavy marine and large stationary applications

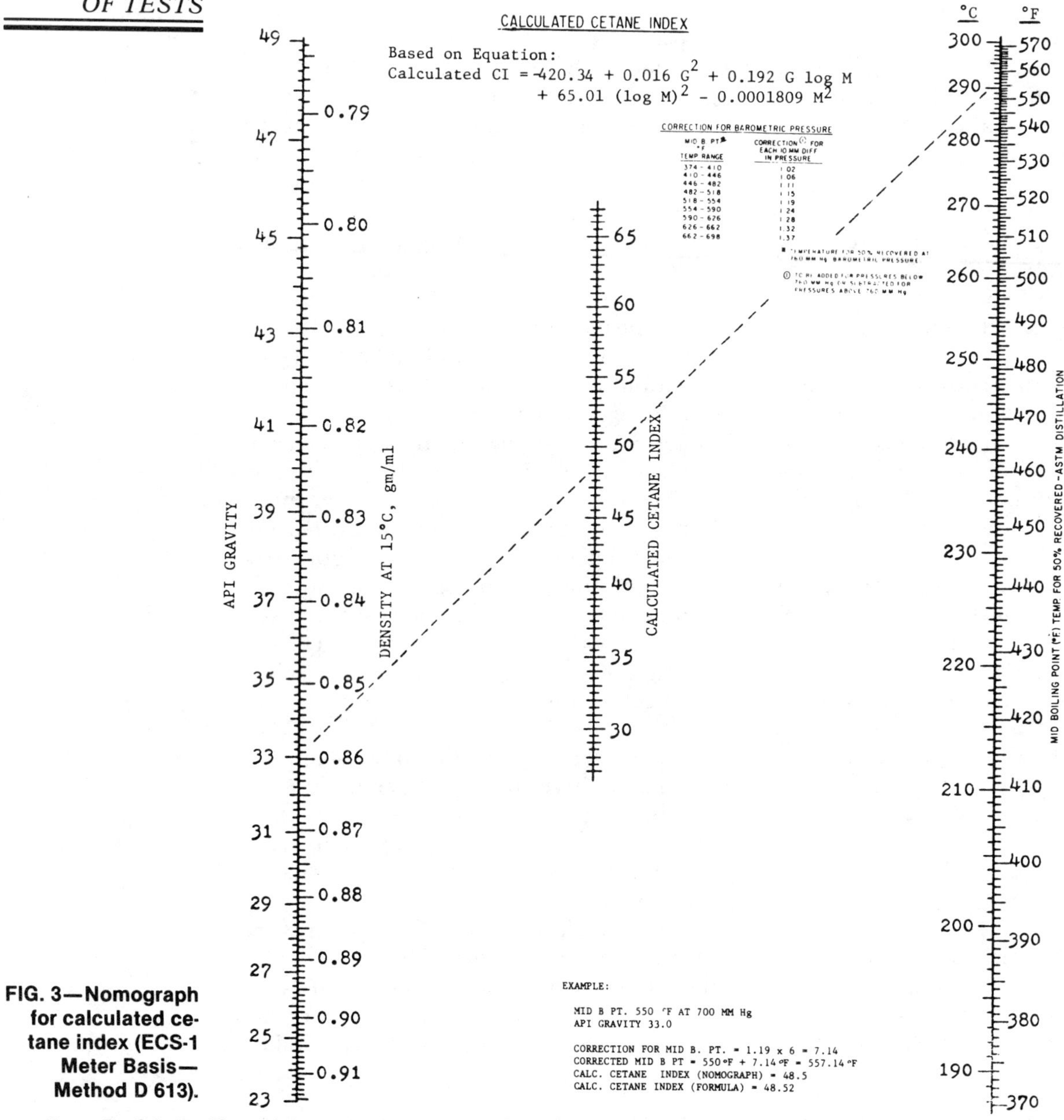

FIG. 3—Nomograph for calculated cetane index (ECS-1 Meter Basis—Method D 613).

NOTE—The Calculated Cetane Index equation represents a useful tool for *estimating* cetane number. Due to inherent limitations in its application, Index values may not be a valid substitute for ASTM Cetane Numbers as determined in a test engine.

residuals (or blends containing residuals), or products having end points below 260°C (500°F).

Diesel Index

The diesel index is derived from the API gravity and ASTM Test for Aniline Point and Mixed Aniline Point of Petroleum Products and Hydrocarbon Solvents (D 611). Aniline point is the lowest temperature at which the fuel is completely miscible with an equal volume of aniline. The formula for diesel index is as follows

Diesel index

$$= \frac{\text{aniline point (°F) API gravity @ 15°C (60°F)}}{100}$$

This equation is seldom used because the results can be misleading, especially when applied to blended fuels.

Gravity

Gravity is an indication of the density or weight per unit volume of a fuel. ASTM D 1298/IP 160 Test Method for Density, Relative Density (Specific Gravity), or API Gravity of Crude Petroleum and Liquid Petroleum Products by Hydrometer Method shows a primary use of specific gravity to convert weights of oil to volumes, or volumes to weights. Specific gravity is also required when calculating the volume of oil at a temperature different from that at which the original volume was measured.

API gravity (ASTM D 1298/IP 160) is an arbitrary figure which permits calculations to be made in whole numbers. It is related to the specific gravity in accordance with the following formula

API gravity, deg

$$= \frac{141.5}{\text{sp gr (15/15°C) (60/60°F)}} - 131.5$$

Where a fuel requires centrifuging, as with marine blended fuels, density is a critical property. As fuel density approaches the density of water, or API gravity of 10 at 15°C (60°F), the efficiency of centrifuging decreases. ISO 8217 limits density to 0.991 kg/m^3 at 15°C (API = 11.2 at 50°C) for most categories of blended marine fuels. Equipment has been developed to function efficiently at API gravities as low as 8.5 at 15°C (60°F) (1010 kg/m^3), but, in general, separators in use are limited to the more traditional 11.2 API gravity.

Where separation, for example, of suspended water from the fuel oil, is not required, specific gravity by itself is not a significant measure of fuel quality. However, when considered with other tests, it may give useful information. For example, for a given volatility range, high specific gravity is associated with aromatic or naphthenic hydrocarbons and low specific gravity with paraffinic hydrocarbons. Further, the specific heat energy a.k.a. the heat of combustion, expressed in energy per unit mass, potentially available from a fuel decreases with an increase in density or specific gravity. However, the heat of combustion expressed per volume of fuel increases with an increase in density or specific gravity.

Distillation

Particularly *in medium- and high-speed engines,* distillation (or volatility) characteristics of a diesel fuel exert a great influence on its performance. Distillation characteristics are measured with a procedure, ASTM Test for Distillation of Petroleum Products (D 86/IP 123), in which a sample of the fuel is distilled and the vapor temperatures recorded for the percentages of evaporation or distillation throughout the range.

The average volatility requirements of diesel fuels vary with engine speed, size, and design. However, fuels having too low volatility tend to reduce power output and fuel economy through poor atomization, while those having too high volatility may reduce power output and fuel economy through vapor lock in the fuel system or inadequate droplet penetration from the nozzle. In general, the distillation range should be as low as possible without adversely affecting the flash point, burning quality, heat content, or viscosity of the fuel. If the 10 percent point is too high, poor starting may result. An excessive boiling range from 10 to 50 percent evaporated may increase warm-up time. A low 50 percent point is desirable to minimize smoke and odor. Low 90 percent and end points tend to ensure low carbon residuals and minimum crankcase dilution.

The temperature for 50 percent evaporated, known as the midboiling point, usually is taken as an overall indication of the fuel distillation characteristics when a single numerical value is used alone. For example, in high-speed engines a 50 percent point above 302°C (575°F) might cause smoke formation, give rise to objectionable odor, cause lubricating oil contamination, and promote engine deposits. At the other extreme, a fuel with excessively low 50 percent point would have too low a viscosity and heat content per unit volume. Therefore, a 50 percent point in the range of 232 to 280°C (450 to 535°F) is desirable for the majority of higher speed type diesel engines. This temperature range usually is broadened for larger, slower speed engines.

The distillation test for *gas-turbine* fuel shows the volatility of the fuel and the ease with which it can be vaporized. A low-volatility fuel results in poor ignition and combustion characteristics. Arctic and antarctic operations, as well as winter conditions, may require additional heat for vaporization or an improved means of atomization for heavier distillates and cold fuels. Distillation temperature is not directly significant to the operation of gas turbines designed for Grade Nos. 3-GT and 4-GT. In gas turbines that are susceptible to carbon deposition and smoke formation, the more volatile fuels may provide better performance.

Viscosity

Viscosity, ASTM Test for Kinematic Viscosity of Transparent and Opaque Liquids (and the Calculation of Dynamic Viscosity) (D 445/IP 71), may be described as a measure of a liquid's resistance to flow. It usually is measured by recording the time required for a given volume of fuel at a constant temperature to flow through a small orifice of standard dimensions. The viscosity of diesel fuel is important primarily because of its effect on the handling of the fuel by the pump and injector system.

In diesel engines, fuel viscosity exerts a strong influence on the shape of the fuel spray. High viscosities can cause poor atomization, large droplets, and high-spray jet penetration. With high viscosities, the jet tends to be a solid stream instead of a spray of small droplets. As a result, the fuel is not distributed in, or mixed with, the air required for burning. This results in poor combustion, accompanied by loss of power and economy. In small engines, the fuel spray may impinge upon the cylinder walls, washing away the lubricating oil film and causing dilution of the crankcase oil. Such a condition contributes to excessive wear.

Low-fuel viscosities result in a spray which is too soft and does not penetrate far enough in the combustion chamber for good mixing. Combustion is impaired and power output and economy are decreased. Low viscosity can lead to excessive leakage past the injection pump plunger. Fuel metering becomes inaccurate and engine efficiency is reduced. Wear of the fuel system components may increase because lubricating properties of fuels tend to decrease with viscosity.

Fuel viscosities for high-speed engines range from 1.8 to 5.8 cSt (32 to 45 SUS) at 37.8°C (100°F). Usually the lower viscosity limit is established to prevent leakage in worn fuel injection equipment as well as to supply lubrication for injection system components in certain types of engines. During operation at low-atmospheric temperatures, the viscosity limit sometimes is reduced to 1.4 cSt (30 SUS) at 37.8°C (100°F) to obtain increased volatility and sufficiently low-pour point. Fuels having viscosities greater than 5.8 cSt (45 SUS) usually are limited in application to the slower-speed engines. The very viscous fuels commonly used in large stationary and marine engines normally require preheating for proper pumping, injection, and atomization.

To produce desired viscosities (30 to 420 cSt at 50°C) in marine diesel fuels, a great deal of blending is frequently required. Either blending of incompatible components or the use of components in near equal volumes or both may produce asphaltine precipitation. This can cause tankage deposits, preheater fouling, and strainer plugging in fuel handling systems, and overloading of centrifuges in marine diesel systems. A paper by Griffin and Siegmund in *Marine Fuels, ASTM STP 878* gives a good account of effective blending procedures.

High-viscosity fuels used in *gas turbines* can result in poor ignition and combustor characteristics and cause excessive pressure losses in the piping system. Even normally free-flowing fuels become thicker, and eventually solid, as temperature decreases. For heavier industrial fuels, fuel temperature must be controlled to assure viscosity suitable for atomization.

Because fuel handling systems rely on the lubrication of pumps and flow dividers by the fuel, very low fuel viscosities are sometimes unsuitable. Fuel systems not needing such lubrication are necessary if very low viscosity fuels are to be used. Alternatively, the use of boundary lubricant additives in the fuel may suffice.

HEAT OF COMBUSTION

The heat of combustion of a fuel is the amount of heat produced when the fuel is burned completely. It may be determined by bomb calorimetric techniques such as those defined by ASTM Test Method for Heat of Combustion of Liquid Hydrocarbon Fuels by Bomb Calorimeter (D 240).

There are two values for the heat of combustion, or calorific value, for every petroleum fuel. They are referred to as the gross and net heats of combustion. The difference between the two calorific values is that the gross heat of combustion includes the heat supplied by the water vapor in condensing, whereas the net heat of combustion does not include this heat. The contribution to the calorific value obtained from other products of combustion, that is, from the nitrogen oxides and the sulfur oxides, are removed from both the gross and the net heats of combustion.

Calorific values can be calculated with an accuracy sufficient for normal purposes from the specific gravity of the product. One relationship that can be used is:

$$Qg = 1.8 \times (12\,400 \text{ to } 2\,100d^2)$$

where

Qg = gross heat of combustion in btu/lb, and
d = specific gravity at 60/60°F (15.6/15.6°C).

The power available from an engine under constant, part-throttle, running conditions and with a constant rate of fuel supply is governed by the calorific value of the fuel. A fuel of low-calorific value yields less heat on combustion and, therefore, less power than the same amount of a fuel with high-calorific value. To maintain power output with the low-calorific value fuel, more of it would have to be used. The importance of this quality depends on whether the user purchases fuel on a weight or volume basis.

Cloud Point

Under low-temperature conditions, paraffinic constituents of a fuel may be precipitated as a wax. The wax settles out and blocks the fuel system lines and filters causing malfunctioning or stalling of the engine. The temperature at which the precipitation occurs depends upon the origin, type, and boiling range of the fuel. The more paraffinic the fuel, the higher the precipitation temperature and the less suitable the fuel for low-temperature operation.

The temperature at which wax is first precipitated from solution can be measured by ASTM Test for Cloud Point of Petroleum Oils (D 2500/IP 219) or by ASTM Test for Wax Appearance Point of Distillate Fuels (D 3117). The cloud point of a fuel is a guide to the temperature at which it may clog filter systems and restrict flow. Cloud point is becoming increasingly important for fuels used in high-speed diesel engines because of the trend toward finer filters. The finer the filter, the more readily it will become clogged by small quantities of precipitated wax. Larger fuel lines, filters of greater capacity, and filters located to receive engine heat reduce the problem and therefore widen the cloud point range of fuels which can be used.

Pour Point

The pour point, ASTM Test for Pour Point of Petroleum Oils (D 97/IP 15), of a fuel is an indication of the lowest temperature at which the fuel can be pumped. Pour points often occur 4.5 to 5.5°C (8 to 10°F) below the cloud points, and differences of 8 to 11°C (15 to 20°F) are not uncommon. Fuels, and in particular waxy fuels, will flow below their tested pour point in some circumstances. However, pour point is a useful guide to the lowest temperature at which a fuel can be used.

No wax precipitation problems are encountered at temperatures above the cloud point and satisfactory operation is unlikely at temperatures below the pour point. The level between these two temperatures at which trouble-free operation is just possible will depend upon the design and layout of the fuel system. A system that contains small, exposed lines and small area fine filters in cold locations will be more prone to early failure than one where the lines are unrestricted and sheltered and where any fine filters are located so that they readily pick up engine heat.

Sometimes additives are used to improve the low-temperature fluidity of diesel fuels. Such additives usually work by modifying the wax crystals so that they are less likely to form a rigid structure. Thus, although there is no alteration of the cloud point, the pour point may be lowered dramatically. Unfortunately, the improvement in engine performance as a rule is less than the improvement in pour point. Consequently, the pour-point temperatures cannot be used to indicate engine performance with any accuracy.

Attempts to develop suitable flow tests have been to date only moderately successful due primarily to the limited amount of firm field operability data on which to base the work. Consequently, there has been a great reluctance to depart from the known and accepted cloud and pour-point tests as the major low-temperature performance criteria. Nevertheles, low-temperature operability tests based on the plugging of cold filters have now been accepted by France (AFNOR-549), Germany (DIN 00 51 770), and Sweden (SIS 155 122). ASTM too has standardized such a test, Test Method for Filterability of Diesel Fuels by the Low Temperature Flow Test (LTFT) Method (D 4539).

Flash Point

The flash point of a fuel is the temperature to which the fuel must be heated to produce an ignitable vapor-air mixture above the liquid fuel when exposed to an open flame. The most common procedure used for determining the flash point of fuels is ASTM Test Methods for Flash Point by Pensky-Martens Closed Tester (D 93/IP 34).

In practice, flash point is important primarily from a fuel handling standpoint. Too low a flash point will cause fuel to be a fire hazard, subject to flashing, and possible continued ignition and explosion. In addition, a low-flash point may indicate contamination by more volatile and explosive fuels such as gasoline. Insurance companies, government agencies, and private users set mandatory minimum limits on flash point because of fire hazard considerations. These limits must be taken into account when establishing fuel specifications. In spite of its importance from a safety standpoint, the flash point of a fuel has no significance to its performance in an engine. Auto-ignition temperature is not influenced generally by variations in flash point nor are other properties, such as fuel injection and combustion performance.

Sulfur

Diesel Engines

Sulfur can cause wear in diesel engines as a result of the corrosive nature of its combustion by-products and increase the amount of deposits in the combustion chamber and on the pistons. The sulfur content of a fuel, ASTM Test for Sulfur in Petroleum Products (General Bomb Method) (D 129/IP 61) and ASTM Test for Sulfur in Petroleum Oils (High Temperature Method) (D 1552), depends on the origin of the crude oil from which the fuel is made and on the refining methods used. Sulfur can be present in a number of forms—as mercaptans, sulfides, disulfides, or heterocyclic compounds such as thiophenes—all of which affect wear and deposits.

Fuel sulfur tolerance by a diesel engine depends largely upon whether the engine is of the low- or high-speed type and the prevalent operating conditions. Low-speed engines can tolerate more sulfur than their high-speed counterparts because they operate under relatively constant speed and load conditions. Under these conditions lubricating oils, cooling water, and combustion zone temperatures show little fluctuation. These steady temperatures make low-speed engines more tolerant to sulfur.

High-sulfur fuels for diesel engines are undesirable from a purely technical standpoint regardless of engine type. However, less harm will occur from fuel sulfur when engines are operated at high-power outputs and operating temperatures than at low temperatures. Under the lower temperature conditions that result from stopping and starting or decrease of load or speed or both, moisture condensation is apt to occur within the engine. The sulfur in the fuel then combines with the water to form acid solutions which corrode metal components and increase wear of moving parts.

Active sulfur in fuel tends to attack and corrode injection system components. Sulfur compounds also contribute to combustion chamber and injection system deposits.

Fuel sulfur is measured both on the basis of quantity and potential corrosivity. The quantitative measurements can be made by means of a combustion bomb (ASTM D 129/IP 61). The measurement of potential corrosivity is determined by means of a corrosion test such as the copper strip procedure described in ASTM Detection of Copper Corrosion from Petroleum Products by the Copper Strip Tarnish Test (D 130/IP 154). The quantitative determination is an indication of the corrosive tendencies of the fuel combustion products while the potential corrosivity indicates the extent of corrosion to be anticipated from the unburned fuel, particularly in the fuel injection system.

It is not unusual for residual type fuels used in the larger, slower-speed engines to have a sulfur content of 3.0 percent by weight or even higher. On the other hand, fuel for high-speed use generally has a sulfur content of 0.4 percent by weight or less to avoid excessive wear. Recommended practices are to maintain the sulfur content as low as practicable.

Modern heavy duty engine oils of high detergency and those containing reserve alkalinity properties minimize the effects of diesel fuel sulfur. However, several factors should be considered before the decision is made to use low-cost, high-sulfur content diesel fuel:

1. Increased cost of higher quality lubricating oil required.

2. Possible increased engine and fuel system wear.

3. Probable fuel system modifications required if a residual fuel is to be considered.

Gas Turbines

Sulfur compounds (notably hydrogen sulfide, elemental sulfur, and polysulfides) can be corrosive in fuel handling systems, and mercaptans can attack any elastomers present. The direct corrosivity of sulfur compounds is measured by the copper strip test, ASTM D 130/IP 30.

Mercaptan sulfur content is limited to low levels because of objectionable odor, adverse effects on certain fuel system elastomers, and corrosiveness toward fuel system metals. Mercaptan sulfur is normally determined by ASTM Test for Mercaptan Sulfur in Aviation Turbine Fuels (Color-Indicator Method) (D 3227/IP 104) or by the Doctor test method, ASTM Specification for Hydrocarbon Drycleaning Solvents (D 235/IP 30).

Sulfur by itself has little corrosive effect on vanes and blades in the turbine section. However, in the presence of alkali metals, sulfur reacts to form alkali sulfates which do promote corrosion at high temperatures. At metal surface temperatures above about 760°C (1400°F), very little sulfur is needed to form the sulfates.

At temperatures between 593°C (~1100°F) and 760°C (1400°F), the concentration of sulfur trioxide (SO_3) in the gaseous combustion products has a far stronger effect because, in order to have corrosion in this temperature range, it is necessary to form sulfates of nickel or cobalt. These compounds, together with the alkalie sulfates, form eutectic mixtures having low melting point temperatures, thus promoting corrosion at the lower temperatures.

In the exhaust section and particularly in installations where waste heat boilers are used, sulfur trioxide together with water vapor can condense as sulfuric acid if the system temperature drops below the acid dew point temperature.

In general, however, the sulfur concentration in a fuel is limited by emission requirements rather than by such technical considerations as those identified above. Methods for determining sulfur are indicated in ASTM Specification for Gas Turbine Fuel Oils (D 2880).

Carbon Residue

The ASTM Test for Conradson Carbon Residue of Petroleum Products (D 189/IP 13) is quoted widely in fuel specifications. The carbon residue is a measure of the carbonaceous material left in a fuel after all the volatile components are vaporized in the absence of air. At one time, there was believed to be a definite correlation between Conradson carbon results in diesel fuels and deposit formation on injector nozzles, but this view now is thought to be an oversimplification.

The type of carbon formed is as important as the amount. Small quantities of hard, abrasive deposits can do more harm than larger amounts of soft, fluffy deposits. The latter can be eliminated largely through the exhaust system.

Carbon residue tests are used primarily on residual fuels since distillate fuels which are satisfactory in other respects do not have high Conradson carbon residue. Because of the considerable difference in Conradson carbon residue results between distillate and residual fuels, the test can be used as an indication of contamination of distillate fuel by residual fuel.

The significance of the Conradson carbon test results also depends on the type of engine in which the fuel is being used. Fuels with up to 12 percent weight Conradson carbon residue have been used successfully in slow-speed engines.

In gas-turbine fuels, carbon residue is a rough approximation of the tendency of a fuel to form carbon deposits in the combustor. Combustion systems designed for use on Grade Nos. 3-GT and 4-GT are insensitive to this problem, but other gas turbines may require a limit on the carbon residue.

Carbon deposits in gas turbines are undesirable because they form heat insulated spots in the combustor that become very hot. Adjacent metal is kept at a relatively low temperature by the cooling air. The "hot spot" formed by carbon deposit creates a large temperature gradient with resultant high stress, distortion, and perhaps eventual cracking of the combustor shell.

Carbon deposits may also contribute to nonuniformity of operation and flow pulsation. If pieces of the carbon deposit are broken off and carried through the turbine, blade erosion (efficiency loss) or partial blocking of the nozzles may occur. Carbon deposits in a combustor usually are accompanied by a smoky discharge and a low value for the heat release factor.

Carbon deposits also occur on the fuel injectors. This disrupts the mixture formation and, consequently, combustion.

Ash

Small amounts of nonburnable material are found in fuels in two forms: (1) solid particles, and (2) oil or water-soluble metallic compounds. The solid particles are for the most part the same material that is designated as sediment in the water and sediment test. These two types of nonburnable material may be oxidized or otherwise modified during the combustion of the fuel.

The quantitative determination for ash is made by ASTM Test for Ash from Petroleum Products (D 482/IP 4). In this test, a small sample of fuel is burned in a weighed container until all of the combustible matter has been consumed. The amount of unburnable residue is the ash content, and it is reported as percent by weight of the fuel.

Since *diesel fuel* injection components are made with great precision to extremely close fits and tolerances, they are very sensitive to any abrasive material in the fuel. Depending on their size, solid particles can contribute to wear in the fuel system and plugging of the fuel filter and fuel nozzle. In addition, abrasive ash materials can cause wear within the engine by increasing the overall deposit level and adversely affecting the nature of the deposits.

The soluble metallic compounds have little or no effect on wear or plugging, but they can contain elements that produce turbine corrosion and deposits as described in later paragraphs.

Gas Turbines

The ash in distillate fuels is typically so low that it does not adversely affect gas turbine performance, unless such corrosive species as sodium, potassium, lead, or vanadium are present. Grade No. 4-GT fuels, however, may have considerable quantities of ash-forming constituents and these may be augmented by the presence of corrosion-inhibiting additives, for example, magnesium, which are used to inhibit corrosion caused by vanadium compounds. In such cases, ash can accumulate on stationary and rotating airfoils, thus restricting gas flows and raising the compressor discharge pressures above the design limits. In addition, the accumulation of ash deposits on rotating airfoils compromises the ability of the turbine to extract work from the expanding combustion gases and thermal efficiency is consequently reduced.

Neutralization Number

ASTM Test for Neutralization Number by Color-Indicator Titration (D 974/IP 139) is a measure of the inorganic and total acidity of the fuel and indicates its tendency to corrode metals which it may contact.

Stability

On leaving the refinery, the fuel will inevitably come into contact with air and water. If the fuel includes unstable components, which may be the case with fuels containing cracked products, storage in the presence of air can lead to the formation of gums and sediments. These gums and sediments can cause filter plugging, combustion chamber deposit formation, and gumming or lacquering of injection system components with resultant sticking and wear.

An accelerated stability test, ASTM Test for Oxidation Stability of Distillate Fuel Oil (Accelerated Method) (D 2274), often is applied to fuels to measure their stability. A sample of fuel is heated for a fixed period at a given temperature, sometimes in the presence of a catalyst metal, and the amount of sediment and gum formed is taken as a measure of the stability.

The thermal stability test measures the tendency of a fuel to form deposits in the fuel system of a gas turbine. In certain types of gas turbines where the fuel system operates at high-fuel temperature, the thermal stability must be specified, as it is for aviation gas turbine fuels.

Water and Sediment

One of the most important characteristics of a fuel is the water and sediment content, ASTM Test Method for Determination of Water and Sediment in Fuel Oils by the Centrifuge Method (D 1796/IP 75). High-water and sediment content is the result of poor handling and storage practices from the time the fuel leaves the refinery until the time it is delivered to the engine injection system.

Water can easily find its way into fuels as a result of breathing in moisture laden air in storage facilities. When sudden drops in atmospheric temperature take place, condensation of moisture occurs. Also, leakage of rain into fuel transportation and storage facilities, leakage of water during shipment by tanker, and the presence of water accumulated in tanks used for storage and handling can cause water contamination.

Sediment generally consists of carbonaceous material, metals, or other inorganic matter. There are several causes of this type of contamination:

1. Rust or dirt present in tanks and lines.

2. Dirt introduced through careless handling practices.

3. Dirt present in the air breathed into the storage facilities with fluctuating atmospheric temperature.

Particularly during storage and handling at elevated temperatures, instability and resultant degradation of the fuel in contact with air contribute to the formation of organic sediment.

Water can contribute to filter blocking and cause corrosion of the injection system components. In addition to clogging of the filters, sediment can cause wear and create deposits both in the injection system and in the engine.

Sediment in blended marine diesel fuels, particularly catalytic fines present in some middle distillates used for blending, is a cause for concern to vessel operators. To improve the information available on the nature of sediments, ASTM developed ASTM Test Method for Inorganic Particles in Marine Residual Fuel Oils by Selective Centrifugal Separation (D 4484). This method covers the determination of inorganic particles, excluding those containing iron, in marine residual fuel oils. In the absence of other contaminants such as sand or dirt, this method provides a gravimetric determination of fluid catalytic cracking process fines in residual fuel oils.

Composition

The chemical composition of a typical fuel is extremely complex because an enormous number of compounds are normally present. It usually is neither practicable nor profitable to perform individual compound analyses. However, it is sometimes helpful to determine the percentages of broad classes of compounds such as aromatics, paraffins, naphthenes, and olefins. A variety of test methods have been proposed. One that has been approved for fuels that boil below 315°C (606°F) is ASTM Test for Hydrocarbon Types in Liquid Petroleum Products by Fluorescent Indicator Adsorption (D 1319/IP 156).

Appearance and Odor

The general appearance, color, and clarity of a distillate fuel are useful controls against contamination by residuals, water, or fine solid particles. Although the small amount of water or solids required to produce an unsatisfactory hazy fuel is usually insufficient to affect the performance of the fuel, customer acceptance is important. Therefore, it is prudent to check by visual inspection that clear fuel is being delivered, for example, by ASTM Test Method for Free Water and Particulate Contamination in Distillate Fuels (Clear and Bright Pass/Fail Procedure) (D 4176).

Similarly, customer acceptance is important with regard to odor, and it is usually politic to ensure the fuel is reasonably free of contaminants, such as mercaptans, which impart unpleasant odors to the fuel.

Vanadium, Sodium, Potassium, Calcium, Lead

Vanadium can form low-melting compounds, such as vanadium pentoxide which melts at 691°C (1275°F) and which causes severe corrosive attack on all of the high-temperature alloys used for gas-turbine blades and diesel engine valves. For example, to reduce the corrosion rate at 871°C (1600°F) on AISI Type 310 steel to a level comparable with the normal oxidation rate, it is necessary to limit vanadium in the fuel to less than 2 ppm. At 10 ppm, the corrosion rate is three times the normal oxidation rate, and at 30 ppm, it is 13 times the normal oxidation rate. However, if there is sufficient magnesium in the fuel, it will combine with the vanadium to form compounds with higher melting points and thus reduce the corrosion rate to an acceptable level. The resulting ash will form deposits in the turbine, but the deposits are self-spalling when the turbine is shut down. For gas turbines operating below 649°C (1200°F), the corrosion of the high-temperature alloys is of minor importance, and the use of a silicon-base additive will further reduce the corrosion rate by absorption and dilution of the vanadium compounds.

Sodium and potassium can combine with vanadium to form eutectics which melt at temperatures as low as 565°C (1050°F) and with sulfur in the fuel to yield sulfates with melting points in the operating range of the gas turbine. These compounds produce severe corrosion, and for turbines operating at gas inlet temperatures above 649°C (1200°F), no additive has been found which successfully controls such corrosion without forming tenacious deposits at the same time. Accordingly, the sodium-plus-potassium level must be limited, but each element is measured separately. Some gas-turbine installations incorporate systems for washing oil with water to reduce the sodium-plus-potassium level. In installations where the fuel is moved by sea transport, the sodium-plus-potassium level should be checked prior to use to ensure that the oil has not become contaminated with sea salt. For gas turbines operating below 649°C (1200°F), the corrosion due to sodium compounds is of minor importance and can be further reduced by silicon-base additives. A high-sodium content is beneficial even in these turbines, because it increases the water-solubility of the deposits and thereby increases the ease with which gas turbines can be water-washed to obtain recovery of the operating performance.

Calcium is not harmful from a corrosion standpoint; in fact, it serves to inhibit the corrosive action of vanadium. However, calcium can lead to hardbonded deposits which are not self-spalling when the gas turbine is shut down and not readily removed by water washing of the turbine. The fuel washing systems used at some gas-turbine installations to reduce the sodium and potassium level also will lower significantly the calcium content of fuel oil.

Lead can cause corrosion, and, in addition, it can spoil the beneficial inhibiting effect of magnesium additives on vanadium corrosion. Since lead only is found rarely in significant quantities in crude oils, its presence in the fuel oil is primarily the result of contamination during processing or transportation.

As a result of these concerns, limits are suggested in the Appendix of ASTM D 2880 Specification for Gas Turbine Fuel Oils (Table 6).

Although the heavier fuels in Grade No. 4-GT, that is, those whose viscosities approach the maximum 638 cSt at 50°C (50 cSt at 100°C) permitted by the specification, are usually washed, inhibited, and analyzed prior to combustion, such practices are seldom practiced with the lighter distillate fuels, such as Grade No. 2-GT fuels. Rather, the distillate fuels are kept usable by practices given in ASTM Practice for the Receipt, Storage, and Handling of Fuels for Gas Turbines (D 4418).

The practices recommended therein attempt to prevent the introduction of contaminant during transportation and storage. In addition, the proper maintenance of fuel storage tanks and the drainage of accumulated water from such storage tanks, can be very effective in maintaining the cleanliness of distillate fuels.

Vanadium levels in *blended marine fuels* are limited in ISO specification 8217 to maximum concentrations of 100 to 600 ppm depending upon the category of fuel.

TABLE 6. Trace metal limits of fuel entering turbine combustor(s).[a]

	Trace Metal Limits, ppm by weight, (max)			
Designation	Vanadium (V)	Sodium plus Potassium (Na + K)	Calcium (Ca)	Lead (Pb)
No. 0-GT	0.5	0.5	0.5	0.5
No. 1-GT	0.5	0.5	0.5	0.5
No. 2-GT	0.5	0.5	0.5	0.5
No. 3-GT	0.5	0.5	0.5	0.5
No. 4-GT			(Consult turbine manufacturers)	

[a]Test Method D 3605 may be used for determination of vanadium, sodium, calcium, and lead.

Applicable ASTM Specifications

D 396	Specification for Fuel Oils
D 975	Specification for Diesel Fuel Oils
D 2880	Specification for Gas Turbine Fuel Oils

Applicable ASTM/IP Standards

ASTM	*IP*	*Title*
D 86	123	Distillation of Petroleum Products
D 93	34	Flash Point by Pensky-Martens Closed Tester
D 97	15	Pour Point of Petroleum Oils
D 129	61	Sulfur in Petroleum Products (General Bomb Method)

Applicable ASTM/IP Standards ***(continued)***

ASTM	*IP*	*Title*
D 130	154	Detection of Copper Corrosion from Petroleum Products by the Copper Strip Tarnish Test
D 189	13	Conradson Carbon Residue of Petroleum Products
D 240	12	Heat of Combustion of Liquid Hydrocarbon Fuels by Bomb Calorimeter
D 445	71	Kinematic Viscosity of Transparent and Opaque Liquids (and the Calculation of Dynamic Viscosity)
D 482	4	Ash from Petroleum Products
D 611	2	Aniline Point and Mixed Aniline Point of Petroleum Products and Hydrocarbon Solvents
D 974	139	Neutralization Number by Color-Indicator Titration
D 976	218	Calculated Cetane Index of Distillate Fuels
D 1298	160	Density, Relative Density (Specific Gravity), or API Gravity of Crude Petroleum and Liquid Petroleum Products by Hydrometer Method
D 1319	156	Hydrocarbon Types in Liquid Petroleum Products by Fluorescent Indicator Adsorption
	63	Sulfur in Petroleum Oils (Quartz-Tube Method)
D 1552		Test Method for Sulfur in Petroleum Products (High-Temperature Method)
D 1796	75	Water and Sediment in Fuel Oils by Centrifuge
D 2274		Oxidation Stability of Distillate Fuel Oil (Accelerated Method)
D 2500	219	Cloud Point of Petroleum Oils
D 3117		Wax Appearance Point of Distillate Fuels
D 3605		Trace Metals in Gas Turbine Fuels by Atomic Absorption and Flame Emission Spectroscopy
D 4418		Practice for Receipt, Storage, and Handling of Fuels for Gas Turbines
D 4484		Inorganic Particles in Marine Residual Fuel Oils by Selective Centrifugal Separation
D 4539		Filterability of Diesel Fuels by the Low Temperature Flow Test (LTFT)

Bibliography

Marine Fuels, ASTM STP 878, C. H. Jones, Ed., ASTM, Philadelphia, 1985.

7

Heating and Power Generation Fuels

INTRODUCTION

ALTHOUGH MOST PETROLEUM PRODUCTS can be utilized as fuels, the term "fuel oil," if used without qualification, may be interpreted differently in various countries. For example, in Europe fuel oil generally is associated with the black, viscous, residual material which remains as the result of refinery distillation of crude oil either alone or in a blend with lighter components, and it is used for steam generation for large, slow-speed diesel engine operation and industrial heating and processing. In the United States, the term "fuel oil" is applied to both the residual type material and the distillate type products such as domestic heating oil, kerosine, and burner fuel oils.

Because fuel oils are complex mixtures of compounds of carbon and hydrogen, they cannot be classified rigidly or defined exactly by chemical formulas or definite physical properties. For purposes of this chapter, the term "fuel oil" will include all petroleum oils heavier than gasoline that are used in burners. Because of the wide variety of petroleum fuel oils, the arbitrary divisions or classifications which have become widely accepted in industry are based more on their application than on their chemical or physical properties. Thus, it is not uncommon to find large variations in properties among petroleum products sold on the market for the same purpose. However, two broad classifications are generally recognized: (1) "distillate" fuel oils and (2) "residual" fuel oils. The latter are often referred to as heavy fuel oils and may contain cutter stock or distillates.

Distillate fuel oils are petroleum fractions that have been vaporized and condensed. They are produced in the refinery by a distillation process in which petroleum is separated into its fractions according to their boiling range. Distillate fuel oils may be produced not only directly from crude oil, that is, "straight-run," but also from subsequent refinery processes such as thermal or catalytic cracking. Domestic heating oils and kerosine are examples of distillate fuel oils.

On the other hand, residual or heavy fuel oils are composed wholly or in part of undistilled petroleum fractions from crude distillation (atmospheric or vacuum), visbreaking, or other refinery operations. The various grades of heavy fuel oils generally are produced to meet definite specifications in order to assure suitability for their intended purpose. Residual oils are classified usually by viscosity in contrast with distillates which normally are defined by boiling range.

Figure 1 depicts the approximate boiling ranges of various distillate fuels in comparison with gasoline. It should be recognized that the designations and uses shown tend to vary in different countries. For convenience, the products will be introduced and discussed under the general headings of kerosine, domestic heating oils, and residual or heavy fuel oils.

Kerosine

Following the discovery of oil in 1859, kerosine became the major petroleum product and was used, initially, for illumination and, subsequently, for heating. While modern technology has diminished the importance of kerosine, it is still utilized as a primary source of light in some lesser developed countries and for standby or emer-

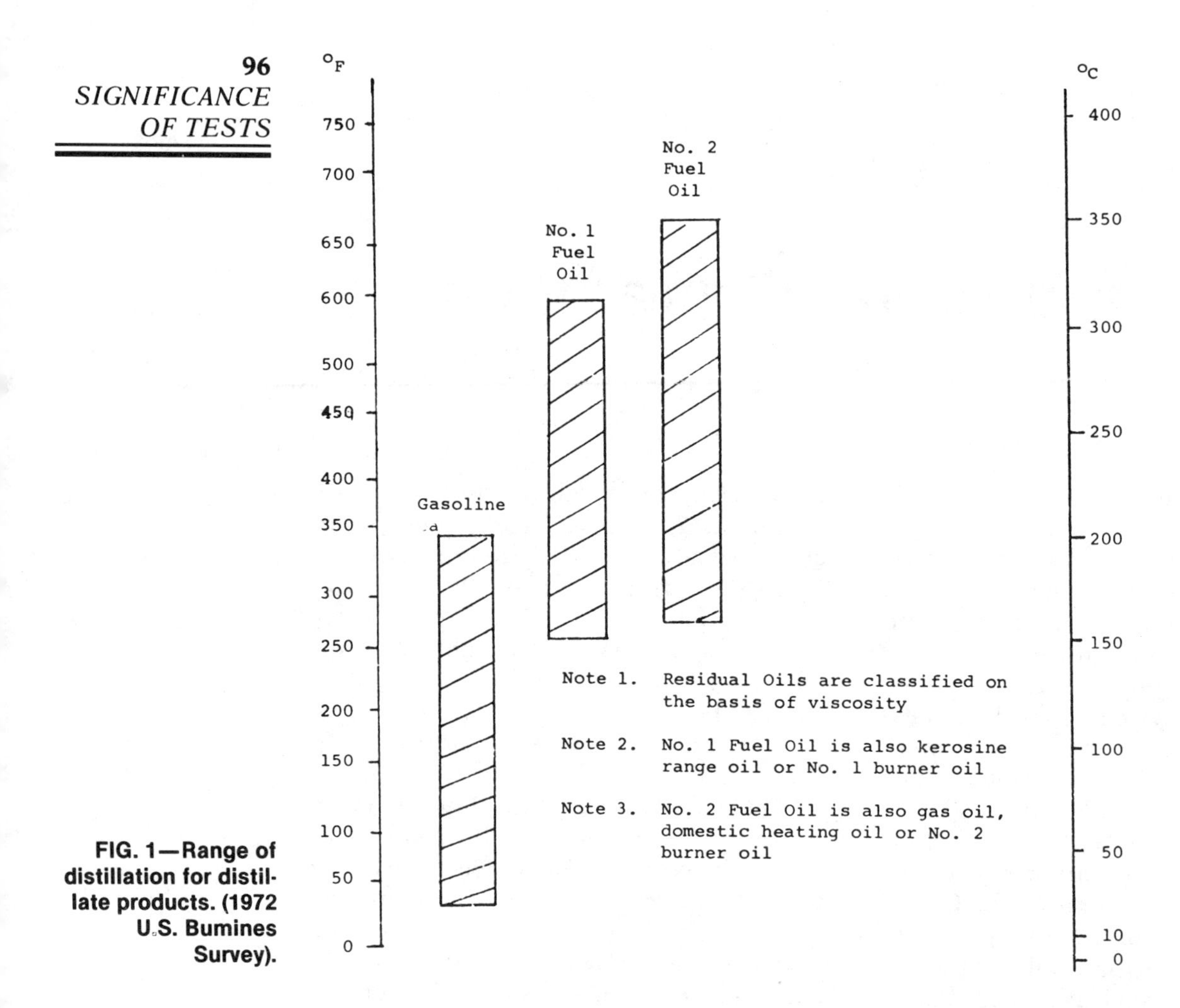

FIG. 1—Range of distillation for distillate products. (1972 U.S. Bumines Survey).

gency lighting in the more advanced areas of the world. Because it is cheap, portable, and flexible, kerosine also is used for domestic space and water heating, refrigeration, heating of garages and greenhouses, incubation and chicken brooders, and cooking. The discussion of kerosine in this chapter will focus on its two primary uses—illumination and heating.

For heating and illumination purposes, kerosine is obtained by fractional distillation of crude oil between approximately 149°C (300°F) and 302°C (575°F) and constitutes a potential 5 to 20 percent volume yield depending on the crude source. (In the United States less than 2 percent of the crude oil is utilized for kerosine.) To decrease smoking, paraffinic stocks are used normally in the manufacture of kerosine for lighting and heating. For the same reason, aromatic stocks and cracked components are avoided.

When low-sulfur paraffinic crudes are fractionated to yield the proper boiling range cut for kerosine, only a drying operation may be required before shipment. Usually, however, some objectionable odors (mercaptans) are present, and these are removed by caustic washing or converted to odorless compounds by sweetening processes.

Kerosine from naphthenic or high-sulfur crudes requires hydrotreating, acid treatment and water wash, or extraction with a solvent and caustic wash and clay brightening to remove undesirable aromatics or sulfur compounds. Following treatment, the kerosine streams are blended to meet specifications, and the finished product is ready for marketing.

Domestic Heating Oils

Domestic heating oils are also distillate products. Their approximate boiling range generally is from 160°C (320°F) to 343°C (650°F). They can include components from three sources: (1) virgin stocks cut directly from crude, (2) oils manufactured by catalytic cracking of heavier stocks, and (3) thermally cracked streams. The percentages of these components blended into the heating oil pool will vary considerably among individual refineries and between various countries. For example, domestic heating oil in most areas outside North America normally consists of straight-run gas oil from the distillation of the crude; in the United States, the straight-run gas oil fraction is usually blended with the appropriate boiling-range material from catalytic cracking process.

With few exceptions, the components cut to the proper boiling range for domestic heating oil require some chemical treatment (for example, additional treatment may be required to remove or convert mercaptans to nonodorous disulfides). Following the chemical treatment, the oils usually are washed with water to remove all traces of the treating residues. As a finishing step, the oil is dried by clay filtration or coalescing.

Domestic heating oils derived from heavier catalytic and thermal stocks may require more severe treatment to remove olefins and other unstable compounds. This is accomplished commonly by catalytic hydrogen treating.

Wherever applicable, such as in the United States, the various heating oil streams finally are blended to meet manufacturing specifications. At this point, a stabilizing inhibitor is often added. To provide the flow properties needed in cold climates, pour-point depressant additives or wax crystal modifiers may be used, or low pour-point kerosine may be blended with the fuel oil.

Residual or Heavy Fuel Oils

Residual fuel oils have a minimum flash point of 60°C (140°F) and viscosities that generally vary between 30 and 300 SFS at 50°C (122°F). The oil viscosity required to obtain satisfactory atomization varies according to the type of burner used and is approximately 40 SFS or 80 cSt for steam atomizing burners and 180 and 200 SUS or 39 to 43 cSt min for mechanical atomizing burners. Therefore, it is necessary to heat the oil in order to achieve the required viscosity reduction. This is done in a heat exchanger, consisting of a tube assembly through which the oil flows, placed inside a shell containing the heating medium. Usually the medium, at a pressure of 20 psig (1.406 kg/cm^2) or less, is steam. Generally, the heating is done in two steps. The first step heats the oil as it is withdrawn from the storage tank, or, in some cases, the whole tank of oil is heated to about 37.8°C (100°F) so that its viscosity is within the pumping range of 400 to 500 SFS or 850 to 100 cSt. In the second step, transfer pumps deliver the oil from storage to the burner heating and pumping equipment at which point it is raised to the required temperature and pressure for atomization.

Crude Oil as a Power Generation Fuel

While crude oil is not typically classified as a fuel oil, attention has begun to focus on its potential as a power generation fuel. Stimulated by the energy shortage, utilities have burned crude oil under power station boilers and as fuel for gas turbines that drive standby generators. Other factors that influenced the utilities' decision to use crude oil in this capacity were environmental considerations, safety, cost, and availability of crude.

The specifications for crude oil as a power generation fuel have not been established, but several known critical properties should be considered. Examples include flash point, sulfur, ash, and metal content. The relatively low-flash points normally associated with crude oils dictate that the storage and handling facilities should be explosion proof, vapor proof, and vapor retaining. To meet stack emissions standards in the United States without resorting to scrubbers or other removal equipment, the crude oil must have low-ash and low-sulfur contents. Low-metal contents are necessary to ensure adequate blade performance when crude is used as a turbine fuel oil.

Additionally, crude oil used as a power generation fuel must meet obvious requirements such as pumpability, and, in the case of turbine use, it must comply with the turbine manufacturers specifications.

The standard tests applied to burner fuels, which will be discussed later in this chapter, are used to test crude oils to be used as fuel. In addition, vapor pressure is determined by the ASTM Test for Vapor Pressure of Petroleum Products (Reid Method) (D 323). Metal contents are measured by the ASTM Test for Trace Metals in Gas Turbine Fuels (Atomic Absorption Method) (D 2788).

BURNING EQUIPMENT

To understand fuel oil tests and their significance, it is necessary to have a fundamental knowledge of the various types of burning equipment associated with kerosine, domestic heating oils, and residual oils. The following paragraphs provide a brief review of the subject.

Kerosine Burners

Although the appliances in which kerosine is used vary widely in design and efficiency, there are three main burner types—wick-fed yellow flame, wick-fed blue flame, and the pressure burner.

The wick-fed yellow flame type burner consists essentially of an oil reservoir into which one end of a wick is immersed. The other end passes through a wick guide and projects upward from it. A draft deflector and a chimney with air inlet are provided. Oil flows by capillary action to the top of the wick where it burns and produces a luminous yellow flame. The flame size may be adjusted by turning the wick up or down, thus exposing more or less area. Either flat or tubular wicks are used. Apart from its domestic use in kerosine lamps and small portable stoves for heating and cooking, the wick-fed yellow flame type burner in simple form is still used in brooders and incubators in the poultry industry. Other applications include miners' safety lamps and certain railway signal lamps.

In the wick-fed blue flame, or kindler, type burner a long drum or chimney is mounted over the burner to induce the air for combustion. The burner itself includes a circular wick and a flame spreader. A well-known application of the wick-fed blue flame method is the perforated sleeve vaporizing burner. This design is known commonly as the "range burner" because of its early widespread use in kitchen ranges. (Kerosine is sometimes referred to as "range oil.")

A typical range burner consists of a flat, cast iron, or pressed steel base with concentric inner-connected grooves and concentric perforated metal sleeves between which combustion takes place. Kerosine is maintained at a depth of 1/8 to 1/4 in. (0.32 to 0.64 cm) in the grooves. Asbestos wicks, lighted manually, are used to provide heat for oil vaporization and ultimate ignition of the oil vapors. As the base heats up, the entering oil vaporizes from the surface, and the flame lights from the wicks. Combustion air is induced by natural draft. The flame is blue, and operation is essentially silent, odorless, and smokeless. A flue usually is provided to remove combustion products. Constant level valves, or other devices such as an inverted bottle based on the "chicken feeder" principle, are required for the kerosine feed.

Appliances of the wick-fed, blue flame type are used primarily for heating and cooking purposes. They also are used with incandescent mantles for lighting. The design is such that intimate mixing of air and oil vapor takes place. This results in more complete combustion, whereby the oil burns with an almost nonluminous blue flame.

Blue flame type kerosine hotwater heaters and vaporizing pot burners will be discussed in the section on domestic heating oil burners, since they can also burn domestic heating oils.

With the pressure burner design, the oil reservoir is fitted with a pump which enables pressure to be maintained. This pressure forces the oil up into a central tube, through a previously heated vaporizing coil, and out through a jet. The issuing oil vapor then mixes with air drawn in from the outside, and the mixture passes to the burner where combustion takes place.

The principle of the pressure-type burner is employed in the Primus stove, various kinds of blow-lamps, certain pressure burners fitted with mantles for illumination, and for a variety of minor industrial applications.

Domestic Heating Oil Burners

For domestic heating oil to burn readily and satisfactorily, it must be vaporized and mixed with the amount of air required to provide a proper combustible mixture. This vaporization can be accomplished either through the application of heat or atomization into very fine particles. Domestic heating oil burners are classified according to the manner in which the vaporization and subsequent burning occurs.

One of the simplest domestic burners is the vaporizing pot burner. These burners are used in room space heaters, water heaters, and for central heating in warm air furnaces. Both natural and mechanical drafts are employed.

A typical pot burner consists of a metal pot perforated with holes for the entrance of combustion air. Oil flows into the bottom of the pot by gravity and is vaporized from the hot surface. After mixing with primary air that enters the lower ports and additional secondary air in the upper section of the pot, the vapors finally burn at the top of the vessel. Between periods of demand or "high-fire" operation, most pot burners idle at a "low-fire" condition. Using only primary air, the flames burn in the bottom section of the pot at a fraction of the high-feed rate and provide the ignition source required for automatic operation.

A typical water heater fired with a natural draft-pot burner requires a relatively high draft [1.5 cm (0.06 in.) of water or more] to provide adequate combustion air. Fuel flow is controlled by the thermostat to give high-fire operation for heating and low-fire operation when the water is at design temperature.

By far the most common of the domestic heating oil burners in the United States and the one fast growing in popularity in other countries is the high-pressure atomizing gun burner. It is the most rugged and among the cheapest of the domestic heating oil burners. In this type burner, oil is supplied to a high-pressure atomizing nozzle, at 5.625 to 8.79 kg/cm^2 (80 to 125 psig.). The high-pressure oil is accelerated in tangential slots to a high velocity. Upon leaving the orifice, the oil breaks into a fine, cone-shaped spray of droplets. Besides the atomizing function, the nozzle also meters the oil.

In the high-pressure gun burner, pump pressure is controlled by a pressure regulating valve which bypasses excess oil to the pump inlet or tank. Combustion air is supplied by a fan mounted on the same shaft as the pump. The air flows through the blast tube in which the nozzle is centered. The ignition electrodes are located slightly behind the spray cone. When power is supplied, the fan and pump begin to rotate and sparking begins between electrodes. The regulating valve opens when atomizing pressure is reached. By this time, the air stream is blowing the center of the spark into the oil spray, and the oil is ignited.

The low-pressure atomizing burner is a more expensive device but offers some advantages over the high-pressure gun. Atomization is accomplished in a nozzle. Oil is supplied at relatively low pressure [0.07 to 1.055 kg/cm^2 (1 to 15 psig.)]. In some cases, a premixed oil and air foam is supplied to this type of nozzle instead of an oil stream. The high velocity of the primary air stream is used to shear off the incoming oil at right angles. A finer spray results than is obtainable from high-pressure nozzles, and this creates a lower smoking tendency. Between 2 and 15 percent of the combustion air is supplied as primary air. Secondary air usually is provided around the nozzle as in the high-pressure gun burner. The oil is metered by means of a metering pump or an orifice in the oil line.

Since there are no oil passages as small as those required for the high-pressure nozzle, the low-pressure type of burner is less prone to plug. This enables the low-pressure burner to function at 1 to 2 L (0.26 to 0.53 gal) an hour while the high-pressure burner, because of a plugging tendency, cannot operate satisfactorily below 2 to 3 L (0.53 to 0.79 gal) an hour.

Vertical atomizing burners produce a flame in the shape of an inverted umbrella and burn in a bowl-shaped hearth. The oil may be atomized from a spinning cup or with a conventional high-pressure nozzle. Sufficient air flow is provided along the walls of the combustion chamber to prevent flame impingement.

The wall-flame rotary vaporizing burner differs radically in principle from the atomizing gun burners. In this equipment, oil is vaporized from a hot metal hearth ring and mixed with air before burning from a set of grills. Its major advantages are higher efficiency and quieter operation. However, the wall-flame rotary burner is more critical of both adjustment and oil quality.

There are various designs of wall-flame rotary burners. In a common type, oil flows by gravity through a metering valve to the central rotor. Centrifugal force throws the oil in a coarsely atomized spray from two rotary distributor tubes to the metal hearth ring. In operation, the ring runs hot enough to vaporize the oil. Combustion air enters the center of the fan and is forced also towards the hearth ring. The air-oil vapor mixture passes upwards through a set of stabilizing grills from which the flame burns.

During start-up, the hearth ring is not hot enough to vaporize the oil. As the rotor starts, oil wets the ring, and sparking begins between the ring and the electrode. A small flame appears at this point and slowly spreads around the hearth ring. As the metal becomes hotter, the flame grows and finally jumps to the top of the grill. Improper adjustment or poor oil quality will cause the formation of deposits on the hearth ring surface. These interfere with cold starting and may result in burner failure.

Residual Oil Burning Equipment

In general, residual oil burning equipment is categorized according to the method by which the fuel is atomized. Described in the following paragraphs are air, steam, mechanical atomizing burners, and rotary type burners.

Air atomizing burners are used largely in industrial furnaces. Because atomizing air mixes intimately with the oil and comprises part of the required combustion air, burning begins more quickly and is completed sooner with this type than with any other type of burner. This results in a shorter flame and permits a smaller combustion space.

Inside-mix air atomizing burners are designed so that all atomizing air enters at the air inlet and is controlled simultaneously with the oil by a single control lever. Primary air enters a whirl chamber tangentially, which imparts a rotating motion to the air. As the air approaches the oil nozzle, it attains maximum velocity because of the venturi shape of the throat. The impact of the air atomizes the oil as it emerges from the nozzle. The resultant air-oil mixture leaves the mixing nozzle in the form of a divergent cone. It comes into contact with the secondary air from an outer nozzle and is further atomized, thus assisting the mixing process.

Steam atomizing burners are two basic types—inside mix and outside mix. As the names suggest, the primary difference lies in where the mixing occurs. In the inside-mix types, the steam and oil mix within the burner nozzle prior to entering the furnace. Additionally, the flame from inside-mix burners may be either flat or conical, depending upon disposition of the burner openings, while the outside mix burners, which were the first type of heavy oil burners, have a flat flame.

Both the outside and inside mix burners are relatively fixed-ratio nozzles which operate best at relatively constant fuel and air supply rates. Because of this limitation, simple inside and outside mix burners are being used less frequently. More sophisticated atomizing burners that yield efficient combustion over a range of firing rates and loads even with very low fuel grades have become available.

In the mechanical atomizing burner, atomization of the fuel oil is achieved by forcing high-pressure, high-velocity oil into the furnace through a small bore orifice or sprayer plate. Required pressure is approximately 21.1 kg/cm^2 (300 psi) or more.

The important part of any mechanical atomizing burner is the sprayer plate. The oil passes through slots in the plate at high velocity. The tangential arrangement of the slots imparts a rotating motion to the oil. The resulting centrifugal force causes the oil to break up into a hollow atomized cone as it enters the furnace through the central orifice.

The firing range of boilers equipped with mechanical atomizing burners can be changed by: (1) changing the number of burners in service, (2) replacing the sprayer plates with plates having another size central orifice, (3) altering the burner design so that more or less oil is allowed to flow through the same size sprayer plate, and (4) changing the oil pressure and method of control in conjunction with a burner design alteration similar to that described in (3).

In the rotary type burners, atomization is achieved through centrifugal force imparted to the oil by a cup rotating at high speed. The atomized oil issues from the rotating cup in the form of a hollow cone. The primary air supply enters the furnace concentrically with this cone and mixes with the oil to form a conical spray. Rapid mixing of the atomized oil and primary air is obtained. The air supplied by the primary air fan is a minor part of the air required for combustion. The secondary air is supplied by either the draft induced by the stack or an additional forced-draft fan and enters the furnace through an air register in the furnace wall.

The rotary burner may consist of a drive motor connected to a shaft that drives the rotating cup, the fuel oil pump and the primary air fan, or the primary air and fuel oil might be supplied by an independently driven unit. The oil viscosity required for the rotating cup burner varies between 150 and 350 SUS (32 to 76 cSt).

FUEL OIL CLASSIFICATION AND SPECIFICATIONS

Because the quality and general performance requirements for various applications of fuel oils differ widely, many countries have adopted general quality limitations for various fuel grades. These serve as guides in the manufacture, sale, and purchase of the oils. While these quality definitions are sometimes called "specifications," they are more properly "classifications" because of their broad, general nature. In addition to these classifications or general specifications, there may be more precise specifications of quality requirements for any given application. These may be dictated by competitive considerations, customer needs, or government agencies.

The general specifications or classifications are usually the least restrictive type and serve primarily as a common basis for agreement between producer, distributor, and consumer. These classifications divide the basic types of oils into broadly defined grades. Broad tolerances permit a customer to select a suitable grade for his purpose, but they do not define the ideal fuel for a particular application. At the same time, the consumer is assured of, at least, minimum performance quality, and there is no undue restriction on the supply of acceptable products.

As an example, the detailed requirements for ASTM Specification for Fuel Oils (D 396) are shown in Table 1. Since ASTM D 396 is revised periodically, the current *Annual Book of ASTM Standards* should be consulted if an up-to-date specification is needed.

FUEL OIL LABORATORY TESTS AND THEIR SIGNIFICANCE

Any property prescribed in a product's specification should be related to product performance or of value in the refining or handling of the product. Specific tests have been designed to determine to what degree a given product meets stated specifications. The test procedures commonly applied to heating and power generation fuels and their significance are summarized in the following paragraphs.

API and Specific Gravity or Density

The gravity of a fuel oil is an index of the weight of a measured volume of the product. There are two scales in use in the petroleum industry: specific gravity—ASTM Test for Density, Relative Density (Specific Gravity), or API Gravity of Crude Petroleum and Liquid Petroleum Products by Hydrometer Method (D 1298/IP 160)—and American Petroleum (API) gravity—ASTM Test for API Gravity of Crude Petroleum and Petroleum Products (Hydrometer Method) (D 287/IP 192).

Specific Gravity of a fuel oil is the ratio of the weight of a given volume of the material at a temperature of 15.6°C (60°F) to the weight of an equal volume of distilled water at the same temperature, both weights being corrected for the bouyancy of air. Specific gravity is seldom used in the United States, but it is in general use in some foreign countries.

API Gravity of a fuel oil is based on an arbitrary hydrometer scale which is related to specific gravity in accordance with the formula

API gravity, deg

$$= \frac{141.5}{\text{sp gr at } 15/15°\text{C } (60/60°\text{F})} - 131.5$$

This scale generally is used for most transactions in the United States as well as in refinery practice.

ASTM Standard D 1250/IP 200 contains tables showing equivalent specific gravity, pounds per gallon, and gallons per pound at 15°C (60°F) for each tenth of a degree API from 0 to 100° API.

Density is the mass (weight in vacuo) of a unit volume of fuel oil at any given temperature [15°C (60°F) as determined in ASTM D 1298/IP 160].

Gravity by itself is of limited significance as an indication of fuel oil quality. It is used by the refiner in the control of refinery operations and has significance to a customer who needs information on net energy release to use in combustion system calculations.

On a weight basis, the heating value of petroleum fuels decreases with increasing specific gravity (decreasing API gravity), since the weight ratio of carbon (low-heating value) to hydrogen (high-heating value) increases as the specific gravity increases. On a volume basis, the increasing specific gravity more than compensates for the decreasing heating value per unit weight, with the net result that fuels having high-specific gravity yield more heat energy per unit volume than those of low-specific gravity.

TABLE 1. Detailed requirements for fuel oils[a] (ASTM D 396).

Grade of Fuel Oil	Flash Point, °F (°C)	Pour Point, °F (°C)	Water and Sediment, volume %	Carbon Residue on 10 % Bottoms, %	Ash, weight %	Distillation Temperatures, °F (°C): 10 % Point	Distillation: 90 % Point		Saybolt Viscosity, s[e]: Universal at 100°F (38°C)		Saybolt: Furol at 122°F (50°C)		Kinematic Viscosity, cSt[e]: At 100°F (38°C)		Kinematic: At 122°F (50°C)		Gravity, °API	Copper Strip Corrosion	Sulfur, %
	Min	Max	Max	Max	Max	Max	Min	Max	Min	Max	Min	Max	Min	Max	Min	Max	Min	Max	Max
No. 1 Distillate oil intended for vaporizing pot-type burners and other burners requiring this grade of fuel	100 or legal (38)	0[c]	0.05	0.15	. . .	420 (215)	. . .	550 (288)	. . .	. . .	. . .	. . .	1.4	2.2	. . .	. . .	35	No. 3	0.5 or legal
No. 2 Distillate oil for general purpose heating for use in burners not requiring No. 1 fuel oil	100 or legal (38)	20[c] (−7)	0.05	0.35	. . .	. . .[d]	540[c] (282)	640 (338)	(32.6)	(37.9)	. . .	. . .	2.0[c]	3.6	. . .	. . .	30	. . .	0.5[b] or legal
No. 4 Preheating not usually required for handling or burning	130 or legal (55)	20[c] (−7)	0.50	. . .	0.10	. . .	. . .	. . .	(45)	(125)	. . .	. . .	5.8	26.4[g]	. . .	. . .	. . .	. . .	legal
No. 5 (Light) Preheating may be required depending on climate and equipment	130 or legal (55)	. . .	1.00	. . .	0.10	. . .	. . .	. . .	(>125)	(300)	. . .	. . .	>26.4[g]	65[g]	. . .	. . .	. . .	. . .	legal
No. 5 (Heavy) Preheating may be required for burning and, in cold climates, may be required for handling	130 or legal (55)	. . .	1.00	. . .	0.10	. . .	. . .	. . .	(>300)	(900)	(23)	(40)	>65	194[g]	(42)	(81)	. . .	. . .	legal
No. 6 Preheating required for burning and handling	140 (60)	. . .[h]	2.00[f]	. . .	. . .	. . .	. . .	. . .	(>900)	(9000)	(>45)	(300)	. . .	. . .	>92	638[g]	. . .	. . .	legal

[a]It is the intent of these classifications that failure to meet any requirement of a given grade does not automatically place an oil in the next lower grade unless in fact it meets all requirements of the lower grade.

[b]In countries outside the United States other sulfur limits may apply.

[c]Lower or higher pour points may be specified whenever required by conditions of storage or use. When pour point less than 0°F is specified, the minimum viscosity for Grade No. 2 shall be 1.8 cSt (32.0 SUS) and the minimum 90 percent point shall be waived.

[d]The 10 percent distillation temperature point may be specified at 440°F (226°C) maximum for use in other than atomizing burners.

[e]Viscosity values in parentheses are for information only and not necessarily limiting.

[f]The amount of water by distillation plus the sediment by extraction shall not exceed 2.00 percent. The amount of sediment by extraction shall not exceed 0.50 percent. A deduction in quantity shall be made for all water and sediment in excess of 1.0 percent.

[g]Where low-sulfur fuel oil is required, fuel oil falling in the viscosity range of a lower numbered grade down to and including No. 4 may be supplied by agreement between purchaser and supplier. The viscosity range of the initial shipment shall be identified and advance notice shall be required when changing from one viscosity range to another. This notice shall be sufficient time to permit the user to make the necessary adjustments.

[h]Where low sulfur fuel oil is required, Grade 6 fuel oil will be classified as low pour (60°F max) or high pour (no max). Low-pour fuel oil should be used unless all tanks and lines are heated.

Flash and Fire Point

The flash point of a fuel is a measure of the temperature to which the fuel must be heated to produce an ignitable vapor-air mixture above the liquid fuel when exposed to an open flame. The "fire point" of a fuel is defined as the temperature at which an oil in an open container gives off vapor at a sufficient rate to continue to burn after a flame is applied. Flash point is normally included in industry specifications for fuel oil.

Flash point is used primarily as an index of fire hazards. As an example, the U.S. Department of Transportation shipping regulations use flash point as the criterion to establish labeling requirements. Consequently, most industry specifications or classifications place limits on the flash point to ensure compliance with fire regulations, insurance, and legal requirements.

A reduction in the normal flash point of an oil may indicate contamination by more volatile products such as gasoline. Therefore, determination of flash point can be useful in detecting such contamination and thereby avoiding serious safety hazards.

Depending upon the apparatus and method of test used, the determined flash point of a fuel may vary. In all cases, however, it is an empirical value because the conditions of the test do not match all possible conditions of commercial handling of the product.

The ASTM Test for Flash Point by Pensky-Martens Closed Tester (D 93/IP 34) and the ASTM Test for Flash Point by Tag Closed Tester (D 56) are normally employed for determining the flash point of fuel oils, because these procedures more nearly approach the conditions of storage of the fuel in tanks.

In addition, a Rapid Flash Test by the Setaflash Tester is receiving increased usage in the United States. Reference to Table 1 will show the flash points specified in the ASTM Standard.

Viscosity

The viscosity of a fluid is a measure of its resistance to flow and is expressed in various units depending upon the equipment and method used for determination. These units include Saybolt Universal seconds, Saybolt Furol seconds, and kinematic viscosity centistokes.

The Saybolt Universal, ASTM Test for Saybolt Viscosity (D 88), and Saybolt Furol, ASTM Conversion of Kinematic Viscosity to Saybolt Universal Viscosity or to Saybolt Furol Viscosity (D 2161), viscosities are used widely in the United States. In the United States, viscosities of the lighter fuel grades are determined by use of the Saybolt Universal viscometer at 37.8°C (100°F) and kinematic viscosity at the same temperature. Saybolt Furol viscosity generally is defined at 50°C (122°F) for the heavier fuels.

The use of these empirical procedures for fuel oils is being superseded by the kinematic system, ASTM Test for Kinematic Viscosity of Transparent and Opaque Liquids (and the Calculation of Dynamic Viscosity) (D 445/IP 71), with a test temperature of 50°C (122°F). ASTM D 445/IP 71 is now shown as the preferred viscosity procedure in ASTM D 396.

The determination of residual fuel oil viscosity is complicated by the fact that some fuel oils containing significant quantities of wax do not behave as simple Newtonian liquids in which the rate of shear is directly proportional to the shearing stress applied. At temperatures in the region of 37.8°C (100°F), residual fuels tend to deposit wax from solution. This wax deposition exerts an adverse effect on the accuracy of the viscosity result unless the test temperature is raised sufficiently high for all wax to remain in solution. Although the present U.S. reference test temperature of 50°C (122°F) is adequate for use with the majority of residual fuel oils, there is some opinion in favor of the higher 82.2°C (180°F) temperature used by the British, particularly in view of the increasing availability of the more waxy fuel oils from the newer North African crudes. A number of difficulties related to anomalous viscosities also would be avoided through adoption of the kinematic system using a test temperature of 82.2°C (180°F).

Viscosity is one of the more important heating oil characteristics. It is indicative of the rate at which the oil will flow in fuel systems and the ease with which it can be atomized in a given type of burner. Since the viscosities of heavier residual fuel oils are high, this property tends to be particularly relevant to its handling and utilization characteristics. The viscosity of a heavy fuel decreases rapidly with increasing temperature. For this reason, heavy fuels can be handled easily and atomized properly by the application of preheat. If no preheating facilities are available, lighter or less viscous oils must be used, and, if the preheating equipment is inadequate, it may be necessary to burn a lighter oil during cold weather.

Overly viscous oil can produce problems throughout the system. Besides being difficult to pump, the burner may be hard to start and flashback or erratic operation may be encountered. Viscosity also affects the output or delivery of a spray nozzle and the angle of spray. With improper viscosity at the burner tip, poor atomization can result in carbonization of the tip, carbon deposition on the walls of the fire box, or other conditions leading to poor combustion.

Typically, the lower limit of viscosity for easy pumpability is reached at about 5000 SUS (1100 cSt). Difficulties in moving the oil in fuel oil systems will be encountered below the temperature at which this viscosity is reached.

For good atomization in pressure and steam atomizing burners, a viscosity range of 100 to 200 SUS (21 to 43 cSt) at the burner is considered desirable. A viscosity at the burner of 80 to 90 SUS (16 to 18 cSt) is recommended frequently with low-pressure air atomizing burners. Rotary cup burners are less critical to atomizing viscosity and can operate satisfactorily with viscosities at the burner cup as high as 350 to 400 SUS (75 to 87 cSt).

Pour Point

Pour point, ASTM Test for Pour Point of Petroleum Oils (D 97/IP 15), is defined as the lowest temperature at which the oil will flow just under standard test conditions. Anticipated storage conditions and fuel application are usually the primary considerations in the establishment of pour-point limits. (Storage of the higher viscosity fuel oils in heated tanks will permit higher pour points than would otherwise be possible.)

While the failure to flow at the pour point normally is attributed to the separation of wax from the fuel, it also can be due to the effect of viscosity in the case of very viscous oils. In addition, pour points, particularly in the case of residual fuels, may be influenced by the previous thermal history of the oils. As an example, any loosely knit wax structure built up on cooling of the oil can be normally broken by the application of relatively little pressure. Thus, the usefulness of the pour-point test in relation to residual fuel oils is open to question, and the tendency to regard pour point as the limiting temperature at which a fuel will flow can be misleading unless correlation is made with low-temperature viscosity.

Although the pour-point test is still included in many specifications, it is not designated for the heavier fuels in ASTM D 396 Grades 5 (light and heavy) and Grade 6. The technical limitations of pour-point have motivated efforts to devise a satisfactory alternative for the pour-point test in assessing the low-temperature pumpability characteristics of heavy fuel oils. Pour-point procedures involving various preheat treatments prior to the pour-point determination and the use of viscosity at low-temperatures have been proposed. However, all of these alternative methods tend to be time consuming and, as such, do not find ready acceptance as routine control tests for assessing low-temperature pumpability or fluidity.

Cloud Point

Distillate fuels, especially, begin to form wax crystals and become cloudy in appearance as they are cooled towards the pour point. The temperature at which this begins to occur is called the cloud point, ASTM Test for Cloud Point of Petroleum Oils (D 2500/IP 219). Cloud points often occur at 4 to 5°C (7 to 9°F) above the pour point, and temperature differentials of 8°C (15°F) are not uncommon. Basically, the temperature differential between cloud and pour point depends upon the nature of the fuel components, but the use of wax crystal modifiers or pour depressants tends to accentuate these differences.

As the temperature continues to decrease below the cloud point, the formation of wax crystals is accelerated. These crystals clog fuel filters and lines and thus reduce the supply of fuel to the burner. Since the cloud point is at a higher temperature than the pour point, it can be considered even more important than the pour point in establishing distillate fuel oil specifications for cold weather usage.

Sediment and Water Content

The tests for water and insoluble solid contents of distillate fuel oils are important to the consumer and among those most commonly applied. Contamination by water and sediment can lead to filter and burner problems and the production of emulsions which are removable only with difficulty. The corrosion of storage tanks also may be associated with water bottoms that accumulate from atmospheric condensation and water contamination.

Water content can be determined by a distillation procedure, ASTM Test for Water in Petroleum Products and Bituminous Materials by Distillation (D 95/IP 74), and sediment can be estimated by a method, ASTM Test for Sediment in Crude and Fuel Oils by Extraction (D 473/IP 53), which involves toluene extraction of the oil through a refractory thimble where the insoluble sediment is retained in the thimble. Total water and sediment can be determined together by a centrifuge procedure, ASTM Test for Water and Sediment in Fuel Oils by Centrifuge (D 1796/IP 75), but separate determinations of water and sediment generally are more accurate.

Ash Content

Ash content may be defined as that carbonaceous, matter-free residue that remains after combustion of the oil in air at a specified, high temperature. To measure ash content, a joint ASTM/IP test method, ASTM Test for Ash from Petroleum Products (D 482/IP 4), is used.

Ash forming materials found in residual fuels are derived normally from the metallic salts found in crude oils. Since crude oil constituents ultimately concentrate in the distillation residue, distillate fuels tend to contain only negligible amounts of ash. However, both distillate and residual fuels may pick up ash contributors during transportation from the refinery. Water transportation, in particular, presents many opportunities for fuel oils to be contaminated with ash producers such as sea water, dirt, scale rust, etc.

The total ash content in different residual fuels is similar normally and less than 0.2 percent by weight. In composition, the ash will vary among residual fuels, largely as a function of their crude oil antecedents, but metallic compounds such as sodium, vanadium, nickel, iron, and silica generally are present.

Depending on the use of the fuel, ash composition has a considerable bearing on whether or not detrimental effects will occur. Ash in heavy fuel oils can cause slagging or deposits and high-temperature corrosion in boilers; it may attack refractories in high-temperature furnaces, kilns, etc., and it may affect the finished product in certain industrial processes such as ceramic and glass manufacture.

Since most of the ash in heavy fuels occurs naturally, it usually is difficult for the refiner to remove it economically. Therefore, methods have been developed for counteracting the effects of ash. These include the use of additives, modifications in equipment design, and the application of fuel processing methods such as water washing.

Carbon Residue

The ASTM Test for Ramsbottom Carbon Residue of Petroleum Products (D 524/IP 14), sometimes called "carbon residue," measures the relative coke-forming propensities or the carbonaceous residue or both and mineral matter remaining after destructive distillation of a fuel oil under certain specified conditions. The carbon residue of burner fuel serves as a rough approximation of the tendency of the fuel to form deposits in vaporizing pot-type and sieve-type burners, where the fuel is vaporized in an air-deficient atmosphere.

Little is known concerning the relationship of carbon residue to fuel performance when applied to residual fuel oils. Pressure jet and steam atomizing type burners are not very sensitive to the carbon residue of the fuel used. In well-designed installations incorporating such burners, and where combustion efficiency is maintained at a high level, it is unlikely that difficulties would normally arise in burning residual fuel oils. Under such circumstances, therefore, it is debatable whether the carbon residue test has any real significance relative to the combustion characteristics of residual fuel.

With distillate fuel oils, high-carbon residue appears to cause rapid carbon buildup and nozzle fouling in certain types of smaller automatic heating units. In the vaporizing pot-type burner described previously, the oil is brought into contact with a hot surface, and the oil vapor subsequently is mixed with combustion air. Any carbonaceous residue formed by the decomposition of the oil or by any incomplete vaporization is deposited in or near the vaporizing surface with resultant loss in burner efficiency. Such a burner, therefore, can be operated satisfactorily only on distillate fuel oils having low-carbon forming tendencies.

To improve the accuracy of carbon residue determination for light distillate fuel oils which form only small amounts of carbonaceous deposits, the carbon residue value is measured on the 10 percent residue obtained by an adaptation of a standard distillation procedure ASTM Test for Distillation of Petroleum Products (D 86/IP 123).

The carbon residue test is, therefore, a useful means for approximating the deposit forming tendencies of distillate type fuel oils used in home heating installations. For this purpose, maximum values for Ramsbottom carbon residue (10 percent residue) are given as 0.15 weight percent for ASTM Grade 1 and 0.35 weight percent for ASTM Grade 2 as shown in Table 1.

Sulfur Content

The sulfur content of fuel oils can be determined by a variety of methods. The ASTM Test for Sulfur in Petroleum Products (General Bomb Method) (D 129/IP 61) and ASTM Test for Sulfur in Petroleum Products (High-Temperature Method) (D 1552) or Quartz-Tube Method (IP 63) have long been established. Other more rapid techniques are also available. These include ASTM Test for Sulfur in Petroleum Products (High-Temperature Method) (D 1552) and ASTM Test for Sulfur in Petroleum Products (X-Ray Spectrographic Method) (D 2622).

Fuel oils contain varying amounts of sulfur depending on the crude source, refining processes, and fuel grade. The high-boiling range fractions and the residual fuels usually contain higher amounts of sulfur which is generally regarded as an undesirable constituent because of its potential to create corrosion and pollution problems.

In boiler systems, the conversion of even a small fraction of the sulfur to sulfur trioxide during combustion of the fuel can cause low-temperature corrosion problems if this gas is allowed to condense and form corrosive sulfuric acid on cool metal surfaces of the equipment. In combination with sodium and vanadium complexes, the sulfur from the fuel contributes to the formation of deposits on external surfaces of superheater tubes, economizers, and air heaters. These deposits cause corrosion of equipment and loss of thermal efficiency.

While desulfurization is being used to reduce the sulfur limits of residual fuel oils, efforts have also been directed towards other means of reducing the effects of acid condensation. One effective procedure is to operate with a minimum of excess air. Another means of reducing low-temperature corrosion is to maintain the cooler metal surfaces of the equipment above the acid dew point; however, this reduces thermal efficiency. Injection of ammonia into the flue gases or the addition of neutralizers are also methods used for this purpose, but these are costly when used in the required concentrations.

Combustion of sulfur-containing fuel oils produces sulfur oxides which have been identified as atmospheric pollutants. To meet clean air standards in densely populated industrial areas, stack emission control devices and sulfur scrubbing procedures may be required.

ASTM D 396 sets maximum sulfur limits for Grades 1 and 2 fuels but does not include any such limits for the heavier grades. Where sulfur content is considered critical, agreement on limits generally is made between the supplier and consumer to meet locally established standards.

When refinery desulfurization processes are used to lower the sulfur content of residual fuels, a lower viscosity fuel (Grade 4 or 5) is usually produced. Also, to meet sulfur limits, fuels of higher pour points 15.6°C (60°F) or higher are being marketed. When the latter fuels are used, adequate preheating facilities are required to assure fluidity and pumpability.

Distillation

The distillation procedures (ASTM D 86/IP 123) measure the amount of liquid vaporized and subsequently condensed as the temperature of the fuel in the distillation flask is raised at a prescribed rate. A record is made of the volume of distillate collected at specified temperatures or, conversely, the temperature at each increment of volume distilled (usually 10 percent increments). The temperature at which the first drop of condensate is collected is called the "initial boiling point." The "end point" usually is the highest temperature recorded as the bottom of the flask becomes dry. If oils are heated above 370°C (700°F), they tend to crack and give unreliable results. The test usually is stopped when this point is reached. Some distillations may be run under a high (10 mm) vacuum in order to avoid cracking, ASTM Test for Distillation of Petroleum Products at Reduced Pressure (D 1160). Under these conditions, heavy fuel oils may be distilled up to temperatures equivalent to 510°C (950°F) at atmospheric pressure. However, distillation tests generally are not applied to the heavy fuel oils.

The distillation test is significant for the distillate fuels. When heating installations that use vaporizing burners or atomizing burners are operated with distillate fuels oils, it is essential that the fuels contain sufficient volatile components to ensure that ignition and flame stability can be accomplished easily. In addition, the volatility of the fuel must be uniform from batch to batch if frequent resetting of burner controls is to be avoided and maximum performance and efficiency are to be maintained.

Table 1 outlines distillation limits for domestic fuel oils—Grades 1 and 2. No distillation limits are established for the other grades of fuels.

Corrosion

Tests for corrosion are of a qualitative type and are made to ascertain whether fuel oils are free of a tendency to corrode copper fuel lines and brass or bronze parts used in the burner assemblies. The method specified for Grade 1 (distillate) burner fuel is ASTM Copper Corrosion from Petroleum Products by the Copper Strip Tarnish Test (D 130/IP 154).

The copper strip corrosion test is conducted by immersing a polished copper strip in a sample of fuel contained in a chemically clean test tube. The tube is then placed in a bath maintained at a temperature of 50°C (122°F) for 3 h. After washing, the strip is then examined for evidence of corrosion and judged by comparisons with the corrosion scale as reflected in the ASTM Copper Strip Corrosion Standards which are lithographed reproductions of copper strips subjected to degrees of corrosion that may be caused by products to which Method D 130 applies.

Additional Tests

Heat Content

The heat content or thermal value of the fuel is the amount of heat given off as a result of its complete combustion. The results are usually expressed as "Kilogram-Calorie" or "British Thermal Unit (Btu)/lb." One kg·cal is equivalent to 1.8 Btu/lb. Heat content determination is made in a bomb calorimeter under specified conditions, ASTM Test for Heat of Combustion of Liquid Hydrocarbon Fuels by Bomb Calorimeter (D 240/IP 12).

The heat content or thermal value test is one which involves careful standardization of equipment and rigid adherence to prescribed details of operation. When properly run, the test can yield highly accurate results. However, it is time consuming and costly to perform on a routine basis.

For normal purposes the heat content or thermal value of a particular fuel can be calculated with sufficient accuracy from other known data. For example, the heat of combustion of distillate fuels has been related to the specific gravity, sulfur content, and aniline point. The heat content of residual fuels is also related to the gravity. (Tables for estimation of net heat of combustion that are applicable to No. 1 burner oil are contained in ASTM Estimation of Net Heat of Combustion of Aviation Fuels (D 1405).)

The heat content or thermal value is not quoted in most fuel oil specifications since it is not directly controllable during manufacture.

Stability

In essence, stability can be defined as the capability of a fuel to resist change in composition. Instability is manifested by a change in color, the formation of gummy materials or insoluble elements, waxy sludge or asphaltic deposition on the bottom of storage tanks, etc.

The storage stability of fuel oils may be influenced by many factors. Among these are crude oil origin, hydrocarbon composition, and refinery treatment. Fuels containing unsaturated hydrocarbons and catalytically cracked components are inherently less stable chemically and have a greater tendency to form sediment on ageing than the straight-run fuels. The presence of reactive compounds of sulfur, nitrogen, and oxygen may also contribute to a fuel oil instability.

Much speculation exists regarding the mechanism of sludge deposition. It may be a consequence of such factors as oxidation, polymerization, and the method of production of the fuel which can result in the formation of insoluble compounds that eventually settle to the tank bottoms and form sludge. The instance of such deposition in light residual fuel oils used in large heating installations may be reflected by clogging of external cold filters, blockage or restriction of pipelines, and combustion difficulties.

Asphaltic deposition may result from the mixing of fuels of different origin and treatment, each of which may be perfectly satisfactory when used alone. Such fuels are said to be incompatible. Straight-run fuels from the same crude oil normally are stable and mutually compatible. Fuels produced from the thermal cracking and visbreaking operations may be stable by themselves but can be unstable or incompatible if blended with straight-run fuels and vice versa.

Instability can be a serious problem that ranges in severity from simple customer dissatisfaction with an off-color fuel to a cause of serious equipment malfunctions. Therefore, test procedures are necessary to predict fuel stability and ensure a satisfactory level of performance by the fuel oil.

A common test is ASTM Test for Stability of Distillate Fuel Oil (Accelerated Method) (D 2274). This method measures the stability of distillate fuel oils under accelerated oxidizing conditions. It should be recognized that any correlation between this test and field storage may vary significantly under different field conditions or with distillates from different sources.

A widely used test to measure the tendency of residual fuel oils to deposit asphaltic matter is U.S. Federal Test Method 3461.1 (NBTL Heater Test). Using this method the fuel oil is circulated over an internally heated steel tube for 20 h at an oil temperature of 93.3°C (200°F). The tube is then examined for asphaltic deposition and rated "stable," "borderline," or "unstable." To assess the compatibility characteristics of the fuel oil, it is blended in equal proportions with each of two reference fuels—one being paraffinic and the other being asphaltic or thermally cracked. Each blend is subjected to the NBTL Heater Test. To be satisfactory in service, the fuel oil by itself must be "stable," and the blended fuels must be assessed as "borderline" or better.

Although no universally accepted procedures are available for assessing the tendency of fuel oils to deposit sludge while in storage, various empirical methods are available within individual petroleum companies.

Maximum Fluidity Temperature

When the determination of fluidity temperature or maximum fluidity temperature of residual fuel oil is required, ASTM Method D 1659 is applicable. This complicated test defines:

1. Fluidity temperature for a given preheating treatment as the lowest temperature to 3°C (5°F) at which a fuel will flow 2 mm in 1 min in a 12.5-mm U-tube under a maximum pressure of 15.2 cm of mercury, and

2. Maximum fluidity temperature as the highest single fluidity temperature found after preheating the fuel to 37, 49, 60, 71, 82, and 93°C (100, 120, 140, 160, 180, and 200°F).

Thermal Stability of U.S. Navy Special Fuel Oil

ASTM Test for Thermal Stability of U.S. Navy Special Fuel Oil (D 1661) covers the determination of the preheated fouling characteristics of fuel oil. The method was developed as an alternate for both the conventional thermal stability and the compatability (NBTL) tests in the U.S. Military Specifications MIL-F-859 for fuel classified as "Burner Fuel Oil—U.S. Navy Special." The test is of 6 h duration and consists of circulation of the fuel sample over the surface of a steel thimble containing a sheathed heating element maintained at 177°C (350°F). At the end of this test period, the thimble is removed and examined for sediment formation and discoloration on the surface.

Spot Tests

To measure compatibility of fuel oil blends, ASTM Test for Compatibility of Fuel Oil Blends by Spot Test (D 2781) is often used. The method covers two spot test procedures for rating residual fuel oil with respect to its compatibility with a specific distillate fuel oil. Procedure "A" indicates the degree of asphaltene deposition that may be expected in blending components. This procedure is used when wax deposition is not considered a fuel application problem. Procedure "B" indicates the degree of wax and asphaltene deposition in the mixture at room temperature. This procedure is used when wax deposition is considered a fuel application problem.

The method is applied to a 50-50 blend of the component fuel oils. A drop of the blend is allowed to spread on chromatographic paper of a specified grade. The spot thus formed is compared with a series of numbered reference spots. Characterizing features of the spots are defined in the method.

DOMESTIC HEATING OIL PERFORMANCE EVALUATION

To evaluate domestic heating oil performance, two ASTM methods are available. Each method provides a means for comparison of burning equipment and for assessment of burning characteristics of the oils. Both can be used either as laboratory or as field procedures.

ASTM Test for Smoke Density in the Flue Gases from Distillate Fuels (D 2156) is used to evaluate smoke density in the flue gases from burning distillate fuels. Since excessive smoke density adversely affects the efficient operation of domestic heating installations, the test is necessary if the smoke production is to be held at an acceptable level.

The second method, ASTM Test for Effect of Air Supply on Smoke Density in Flue Gases from Burning Distillate Fuels (D 2157), is applicable only to equipment incorporating pressure atomizing and rotary wall-flame burners. In this method, efficiency of operation is related to clean burning of the fuel. The extent to which combustion air can be reduced without producing an unacceptable level of smoke is indicative of the maximum efficiency for a particular installation at any acceptable smoke level.

KEROSINE PERFORMANCE EVALUATION

Selection of Burner

Although the performance of a particular fuel oil depends to some extent on its inherent properties, with normal kerosine, performance is more dependent on burner design than on hydrocarbon type composition. Consequently, selection of a burner in which to assess the burning quality of an oil assumes great importance.

For kerosine performance evaluation, simple wick-fed yellow flame burners are used because kerosine type and quality are more critical in these units. Kerosines which are mainly paraffinic burn well in wick-fed yellow flame lamps with a poor draft, while kerosine containing high proportions of aromatics and naphthenes burn with a reddish or even smoky flame under the same conditions. Predominantly aromatic kerosines can burn brilliantly in a wick-fed blue flame lamp with good draft characteristics, while the paraffinic type may burn with a flame of comparatively low-illuminating value.

Since combustion is more complete in the wick-fed blue flame burner than in the yellow-flame type, the former shows less differentiation between kerosine types. Even less differences between oil types are exhibited with pressure burners, which can operate satisfactorily with a wide range of kerosines.

Burning Characteristics

Initial Flame Height Smoke Point

As the wick of a yellow-flame type lamp is turned up, a point is reached where smoking occurs. Therefore, the degree of illumination possible depends mainly on the height of the nonsmoking flame obtainable. This height varies according to the hydrocarbon type composition of the kerosine. The maximum height of flame obtainable without smoking, termed the "smoke point," is greatest with paraffins, considerably lower with naphthenes, and much lower still with aromatics.

Even if full advantage is not taken to utilize maximum nonsmoking flame heights, the characteristics of high-smoke point ensures that there will be less tendency for smoking to occur in the event that a sudden draft causes extension in flame height.

Although a low-smoke point is undesirable, in that it may not yield a satisfactory range of smokeless performance, a high-smoke point alone is no guarantee that a kerosine has generally satisfactory burning characteristics.

ASTM Test for Smoke Point of Aviation Turbine Fuels (D 1322) is the only test for the determination of smoke point used in the United States. This method consists of burning a sample in an enclosed lamp with scale. The maximum flame height that can be achieved without smoking is estimated to the nearest 0.5 mm.

Constancy of Feed to Wick

The maintenance of the initial degree of illumination in a lamp depends on the constancy of flow of the kerosine to the wick as well as on the conditions of the wick. The quantity of oil flowing up a wick is related to the height of the top of the wick above the level of oil in the container and the viscosity and surface tension of the oil. Viscosity is most significant in this respect in surface tension because viscosity varies more in magnitude with different kerosines and changes in temperature.

When a kerosine warms-up during the initial burning period, the flame size tends to increase slightly as a result of increase in evaporation rate and decrease in viscosity. In the case of lamps not provided with constant level feed, the viscosity becomes significant, since the height of unimmersed wick increases with the consumption of the kerosine. If the viscosity is too high, the feed of kerosine and, consequently, the flame height and stability can be reduced seriously. The presence of moisture in a wick also hinders the upward flow of kerosine and causes a drop in flame height and a decrease in kerosine consumption.

Formation of Char on Wick

After a kerosine has been burning for some time, the condition of the exposed wick of a lamp begins to be affected by the formation of a carbonaceous incrustation or char. This is not significant unless it affects the flame or the mechanism for adjusting the wick.

The char may be either hard and brittle or soft, and the amount and appearance will vary according to the nature and properties of the kerosine burned. It may be of irregular formation, which causes flame distortion by producing localized deposits known as "mushrooms" on the surface of the wick, or it may be formed in such an amount and manner as to restrict the size of the flame and, in serious cases, extinguish it.

The formation of char depends mainly on the chemical composition and purity of the kerosine and can be affected by the nature of the wick and the design and operating conditions of the lamp. Possible causes of high-char formation may be:

1. Insufficient refining, resulting in the presence of deleterious impurities in the oil.

2. The presence of high-boiling residues that do not vaporize easily on the top of the wick and thereby cause decomposition and carbonization. Contamination with even minute amounts of heavier products, such as lubricants or fuel oil, can cause serious high-char formation.

Formation of Lamp-Glass Deposits

There should not be any appreciable formation of deposits or "bloom" on the lamp-glass chimney during burning under normal operating conditions. Such bloom, when it does appear, may either be white, gray, brown, or blue in color and should not be confused with the brownish-black or black deposits caused by a smoky flame.

Certain factors such as the design, composition, temperature of the glass chimney, and the purity of the atmosphere can have a bearing on bloom formation. However, this formation is due primarily to the deposition of sulfur compounds derived from the sulfur content of the oil.

Standard Burning Tests

ASTM Test for Burning Quality of Kerosine (D 187/IP 10) provides a good assessment of the burning characteristics discussed previously with the exception of maximum flame height. The tests are applied generally to kerosine used as an illuminant or as a fuel for space heaters, cookers, incubators, etc.

In both methods, the oil is burned for 24 h in a standard lamp with a flame initially adjusted to specified dimensions. The details of operations are carefully stated and cover test-room conditions, volume of sample, wick nature, pretreatment of the wick and glass chimney, method of wick trimming, and the removal of the char.

At the conclusion of the tests, the kerosine consumption is calculated, and a qualitative assessment of the appearance of the glass chimney is made. In the IP method, the amount of char formed on the wick is determined, and the char value is calculated as milligrams per kilogram of kerosine consumed. The ASTM method utilizes only qualitative assessment of the wick, and no quantitative determination of char value is made.

Aside from the intrinsic significance of char value with respect to oil quality, real differences in such values in a series of kerosines enable a relative comparison of burning quality to be made. Even traces of high-boiling contaminants greatly increase the char-forming tendencies of kerosines.

For an assessment of the maximum flame height at which a kerosine will burn, a Smoke Point Test (IP 57) may be used. In this test, a sample is burned in a special lamp in which flame height is varied against the background of a graduated scale and adjusted until the smoky tail of the flame just disappears.

Burning Test for Long-Time Burning Oils

Burning tests for long-time burning oils are used to evaluate the burning properties of oils for use in railway signal lamps and similar kinds of lamps for which long periods of uninterrupted and unattended burning may be necessary.

In Burning Test—7 Day, (IP 11), the oil is burned for 7 days in one of two specified signal lamps under strictly controlled conditions. The flame height and condition of the lamp are recorded every 24 h during the test. At the end of the test, the total consumption and, if required, the char value are determined.

Applicable ASTM Specification

D 396 Specification for Fuel Oils

Applicable ASTM/IP Standards

ASTM	*IP*	*Title*
	11	Burning Test—7-Day
D 56		Flash Point by Tag Closed Tester
	57	Smoke Point
D 86	123	Distillation of Petroleum Products
D 88		Saybolt Viscosity
D 93	34	Flash Point by Pensky-Martens Closed Tester
D 95	74	Water in Petroleum Products and Bituminous Materials by Distillation
D 97	15	Pour Point of Petroleum Oils
D 129	61	Sulfur in Petroleum Products by the Bomb Method
D 130	54	Detection of Copper Corrosion from Petroleum Products by the Copper Strip Tarnish Test
D 187	10	Burning Quality of Kerosine
D 240	12	Heat of Combustion of Liquid Hydrocarbon Fuels by Bomb Calorimeter
D 287	192	API Gravity of Crude Petroleum and Petroleum Products (Hydrometer Method)
D 445	71	Kinematic Viscosity of Transparent and Opaque Liquids (and the calculation of dynamic viscosity)
D 473	53	Sediment in Crude and Fuel Oils by Extraction
D 482	4	Ash from Petroleum Products
D 524	14	Ramsbottom Carbon Residue of Petroleum Products
D 1160		Test for Distillation of Petroleum Products at Reduced Pressures
D 1250	200	Petroleum Measurement Tables
D 1298	160	Density, Relative Density (Specific Gravity), or API Gravity of Crude Petroleum and Liquid Petroleum Products by Hydrometer Method
D 1322		Smoke Point of Aviation Turbine Fuels
D 1405		Estimation of Net Heat of Combustion of Aviation Fuels
D 1552		Sulfur in Petroleum Products (High-Temperature Method)
	63	Sulfur in Petroleum Oils (Quartz-Tube Method)
D 1659		Maximum Fluidity Temperature of Residual Fuel Oil
D 1661		Thermal Stability of U.S. Navy Special Fuel Oil
D 1796		Determination of Water and Sediment in Fuel Oils by Centrifuge Method (Laboratory Procedure)
D 2156		Smoke Density in the Flue Gases from Burning Distillate Fuels
D 2157		Effect of Air Supply on Smoke Density in Flue Gases from Burning Distillate Fuel
D 2161		Conversion of Kinematic Viscosity to Saybolt

Applicable ASTM/IP Standards *(continued)*

ASTM	*IP*	*Title*
		Universal Viscosity or to Saybolt Furol Viscosity
D 2274		Test for Oxidation Stability of Distillate Fuel Oil (Accelerated Method)
D 2500	219	Cloud Point of Petroleum Oils
D 2622		Sulfur in Petroleum Products (X-Ray Spectrographic Method)
D 2781		Compatibility of Fuel Oil Blends by Spot Test

Bibliography

Francis, W., *Fuels and Fuel Technology*, 2 Vols., Bergman Press, Oxford, England, 1965.

Kewley, J. and Gilbert, C. L., *Kerosine*, Vol. IV, Science of Petroleum, 1938.

Modern Petroleum Technology, Institute of Petroleum, London, England, 3rd edition, 1962.

Petroleum Products Handbook, V. B. Guthrie, Ed., McGraw-Hill, New York, 1960.

8

Lubricating Oils

INTRODUCTION

THE MAJOR FUNCTION OF LUBRICATING OILS is the reduction of friction and wear by the separation of surfaces, metallic or plastic, which are moving with respect to each other. The oils also act as carriers for many special chemicals such as corrosion inhibitors, antiwear agents, load-carrying friction modifiers, and foam suppressors. Performance requirements can also include cooling and the dispersion and neutralization of combustion products from fuels. The high quality and improved properties of present-day lubricants have enabled engineers to design machines with higher power-to-weight ratios which generally have higher stresses, loads, and operating temperatures than before. Thus, it has been possible to develop automobile engines, turbines, gear sets, etc. capable of higher speeds and higher specific power output per pound of machinery. In a very different field, lubricants with increased resistance to the effects of radiation have been developed for nuclear power stations.

New base oil refining methods yield stock oils which are more responsive to additive treatment. Research in the field of additives has, in turn, produced lubricant formulations which can operate under the higher piston-ring belt temperatures of super-charged automotive diesel engines and provide the dispersion required to prevent the formation of low-temperature sludges in gasoline engines for stop-start, short distance motoring.

In spite of the increasing temperatures, loads, and other requirements imposed on lubricants, mineral oils are likely to continue to be employed in the foreseeable future for the majority of automotive, industrial, and marine applications. However, in the aviation field, synthetic lubricants are used extensively. There are also a growing number of critical automotive, industrial, and marine applications where the use of synthetic lubricants can be justified on the basis of total performance cost or fire resistance.

Assessment of Quality

New uses and formulation technology for lubricating oils necessitate a constant review of the methods for assessing the quality of both new and used lubricating oils. The traditional physical and chemical tests are still applied, but these are being supplemented and, in some cases, replaced by instrumental techniques based on physico-chemical methods which include infrared absorption, ultraviolet absorption, emission spectroscopy, X-ray absorption, and fluorescence methods. It is convenient to consider these tests in five categories. The first three determine the characteristics and compositions of lubricants, and the last two are laboratory tests that simulate service conditions.

Physical Tests—Physical tests are comparatively simple laboratory procedures which define the nature of the product by measuring physical properties. Examples are viscosity, flash point, specific gravity, color, and appearance.

Chemical Tests—Chemical tests define the composition of the lubricating oil by determining the presence of such elements as sulfur, chlorine, phosphorus, and metals which often are related significantly to the additive content of the lubricant.

Physico-Chemical Tests—Tests in this classification are either (1) those which determine the presence of elements using instrumented physical procedures, (2) those which give information on the molecular structure of the components of the lubricant, or (3) those which give pH, acidity, and alkalinity values.

Physico-chemical tests are used generally to characterize products for specific applications, provide quality control at blending plants, and check the suitability of used lubricants for further service.

Laboratory Bench Tests—Laboratory bench tests subject the lubricating oil to individual environmental conditions which are designed normally to exceed the appropriate service requirements. These include such glassware tests as thermal stability, oxidation stability, and corrosion resistance which are used frequently for screening formulations during the development of a new product.

Also falling in this category are test machines for measuring wear and load-carrying properties. Examples are the 4-Ball, Timken, FZG, and Falex machines.

Engine and Rig Tests—Mechanical tests are used to assess the effects on various properties of lubricating oils that will be produced by the environment in which the lubricant will function. The equipment is set up in a prescribed manner on laboratory stands, and tests are carried out under carefully controlled conditions. Such tests generally are designed to correlate as far as is possible with actual service, but, for new products, laboratory mechanical testing usually is followed by field evaluation.

Significance of Tests

Physical tests, chemical tests, laboratory bench tests, and engine tests are extremely valuable as tools for attempting to predict how a specific lubricant formula will perform in full size machinery under many different operating conditions. They must always be used and interpreted, however, with the full realization that they are not infallible or foolproof. The ultimate decision as to the success or failure of a lubricating fluid can be made only on the basis of its behavior in the end-use item such as production engines, pumps, gear drives, hydraulic systems, etc. It is usually on the basis of this ultimate performance that many equipment manufacturers will "recommend" a brand name product for use in their equipment.

COMPOSITION AND MANUFACTURE

Petroleum base lubricating oils are present in the residue boiling above 370°C (698°F) from the atmospheric distillation of selected crude oils of both paraffinic and naphthenic types. This residue is further distilled, under conditions of high vacuum, into a series of fractions to provide light to heavy lubricating oil stocks. The number of fractions depends on the type of crude oil and the requirements of the refiner, but four to five is a typical number. These basic stock oils are further refined, usually by solvent extraction and solvent dewaxing but sometimes by hydrofinishing, to produce oils suitable for incorporation into finished lubricants. The individual refined stock oils from one or more crude sources are blended in various proportions to provide lubricating oils suitable for a wide range of applications. The blending process can be by mechanical or air agitation and can be either by a batch or continuous in-line method.

Only for the less severe uses is it possible to employ nonadditive mineral oils. In the majority of cases to meet specific applicational requirements, chemical additives are used to enhance the properties of base oils. Additives are used to improve such characteristics as oxidation resistance, change in viscosity with temperature, low-temperature flow properties, corrosion and radiation resistance, and load-carrying capacity. Lubricants frequently contain a number of additives to achieve a balance of properties suitable for the intended application. These must be compatible with the base oils, other additives present, and additives which are commonly used by others who manufacture products intended for similar uses. Thus, the proper selection of the components for this lubricating oil formulation requires knowledge of the most suitable crude sources for the base oils, the type of refining required, the types of additives necessary, and the possible interactions of these components on the properties of the finished lubricating oil.

Control of product quality at the blending plant is usually based on a supplier's own internal standards. The number of tests applied varies with the complexity of the product and the nature of the application. The more important tests (viscosity, flash point, color, etc.) usually are performed on every batch. Other tests may be on a statistical basis dependent on data developed at the individual blending plant. A newer method of control testing includes infrared spectroscopic analysis which can be presented in the graphical form of a "fingerprint" and is specific for the blend of mineral oils and additives in a particular formulation. Comparison of the "fingerprint" with a known standard can be used as a check on the composition.

GENERAL PROPERTIES

Before describing the quality criteria for some of the more important types of lubricating oils, it will be fruitful to discuss general properties common to most lubricating oils and the methods used to determine these properties. In this discussion, pertinent American Society for Testing and Materials and Institute of Petroleum (ASTM/IP) test methods will be listed.

Viscosity

The viscosity of a lubricating oil is a measure of its flow characteristics. To meet a particular application, viscosity is generally the most important controlling property for manufacture and selection. While the viscosity of a mineral oil changes with temperature, it usually does not change with shear-rate, unless specific nonshear stable additives are used to modify viscosity/temperature characteristics—an aspect that is discussed more fully in the section on Automotive Engine Oils.

At very high pressures (several thousand psi) the viscosity of mineral oils increases considerably with increase in pressure. The extent of the viscosity change depends on the crude source of the oil and on the molecular weight of the constituent components. Kinematic viscosity is measured by timing the flow of a fixed amount of oil through a calibrated glass capillary tube under gravitational force at a standard temperature. ASTM Test for Kinematic Viscosity of Transparent and Opaque Liquids (and the Calculation of Dynamic Viscosity) (D 445/IP 71). The unit of viscosity used in conjunction with this method is the centistoke. This unit may be converted into the other viscosity systems (Saybolt, Redwood, Engler) by means of suitable tables.

Because the main objective of lubrication is to provide a film between load-bearing surfaces, the selection of the correct viscosity for the oil is aimed at a balance between a viscosity high enough to prevent the lubricated surfaces from contacting and low enough to minimize energy losses through excessive heat generation caused by having too viscous a lubricant.

The "classical" hydrodynamic theory for moderately loaded bearings predicts complete separation between metallic surfaces with a comparatively thick layer of fluid oil, while highly loaded gears are considered to be in a state of boundary lubrication in which opposing surface irregularities cause metal-to-metal contact to occur. Modern elastohydrodynamic theory for lubricated surfaces takes into account that, because of the high pressure generated, the viscosity of the oil increases considerably and elastic deformation of the surfaces occur. Under these conditions, it has been shown that the lubricant film approaching "boundary" conditions is thicker than previously was supposed.

The viscosity of a new oil is of fundamental importance with respect to performance in a specific type of equipment or machine element and always described or specified by the buyer, the seller, or both.

The Society of Automotive Engineers (SAE) numbers (10W, 20, 30, 40, etc.) are well-known, widely used, and almost universally accepted as a concise but satisfactory way of describing the viscosity characteristics of oils used in the crankcase or gear drive of automotive equipment.

Describing the required, or desired, viscosity characteristics of an oil to be used in the bearing, gears, or hydraulics of industrial machinery or equipment has proven to be much more complicated than the simple use of the SAE numbers. Many differing reference temperatures have been employed in the many viscosity units (Saybolt Universal Seconds, Redwood, Degree Engler) have been promoted, but there has been little or no agreement as to the viscosity limits or the number of grades which are needed by industry. Since 1968, however, there has been a strong move underway by the leading lubricant producers, major machinery manufacturers, and large consumers to adopt a uniform practice. In that year the ASTM and the American Society of Lubrication Engineers (ASLE) Recommended Practice for Viscosity System for Industrial Fluid Lubricants was published as ASTM Standard D 2422. Simultaneously, the British Standards Institute published an identical standard (BS-4231). These two standards recommend a series of viscosity grades, each being approximately 50 percent more viscous than its preceding grade. Both standards described the viscosity of each grade in centistokes at 37.8°C (100°F) and established an allowable deviation of plus or minus 10 percent from the nominal. The use of this uniform system proved to be satisfactory in the United States and the United Kingdom, but it was somewhat out of step with the rest of the world. Consequently, in 1972 efforts by Technical Committee 28 of the International Standardization Organization produced agreement on a reference temperature of 40°C (104°F).

Viscosity increase in a used oil, that is, an oil in service, usually indicates that the oil has deteriorated by oxidation or contamination, while a decrease usually indicates dilution by a lower viscosity oil or a fuel. Viscosity blending charts may be used to estimate the amount of dilution. The extent of the viscosity change permissible before corrective action is required differs in various applications.

The ASTM Viscosity-Temperature Charts for Liquid Petroleum Products (D 341) are useful for estimating the viscosity of an oil at the various temperatures which are likely to be encountered in service.

Viscosity Index

The viscosity of petroleum base oils decreases with a rise in temperature, but this rate of change depends on the composition of the oil. The viscosity index is an empirical number which indicates the effect of change of temperature on the viscosity of an oil. It compares the rate of change of viscosity of the sample with the rates of change of two types of oil having the highest and lowest viscosity indices at the time (1929) when the viscosity index scale was first introduced. A standard paraffinic oil was given a viscosity index (VI) of 100 and a standard naphthenic oil a VI of 0. Equations were evolved connecting the viscosity and temperature for these two types of oil, and, from these equations, tables were prepared showing the relationship between viscosities at 37.8°C (100°F) and 98.9°C (210°F) for oils with a VI between 0 and 100. With these tables and the viscosities at 100 and 98.9°C (210°F) of an oil, the viscosity index can be calculated. A high-viscosity index denotes a low rate of change of viscosity with temperature.

The use of additives and modern refining techniques allows oils to be produced with high-viscosity indices regardless of the type of crude oil from which they originated. At the same time, viscosity improvement additives can produce oils with viscosity indices greater than 100. Initially, this problem was solved by simply extrapolating the original tables, but, as VIs rose ever higher, this produced anomalies.

In 1964, ASTM adopted an extension to the tables based on an equation developed for the purpose. Values derived from this equation are designated VI_E to distinguish them from the original VI. This method has since been adopted by IP under the joint designation ASTM Calculating Viscosity Index from Kinematic Viscosity at 40 and 100°C (D 2270/IP 226). This replaces the former method ASTM D 567/IP 73. (Viscosity indices below 100 have not been affected by this revision.)

The viscosity index of an oil is of importance in applications where an appreciable change in the temperature of the lubricating oil could affect the start-up or operating characteristics of the equipment. The automatic transmission for passenger vehicles is an example of equipment where high-viscosity index oils using VI improvers are employed to minimize differences between a viscosity low enough to permit a sufficiently rapid gear shift when starting under cold conditions and a viscosity adequate at the higher temperatures encountered in normal running.

Cloud and Pour Points

Petroleum oils contain components with a wide range of molecular sizes and configurations and thus do not have a sharp freezing point. They become semiplastic solids when cooled to sufficiently low temperatures.

The cloud point of a lubricating oil is the temperature at which paraffinic wax and other readily solidifiable components begin to crystallize out and separate from the oil under prescribed test conditions, ASTM Test for Cloud Point of Petroleum Oils (D 2500/IP 219). Cloud point is of importance when narrow clearances might be restricted by accumulation of solid material (for example suction line strainers, small size oil-feed lines or filters).

The pour point is the lowest temperature at which the oil will just flow under specified test conditions, ASTM Test for Pour Point of Petroleum Oils (D 97/IP 15), and is roughly equivalent to the tendency of an oil to cease to flow from a gravity-fed system or container. Since the size and shape of the containers, the head of the oil, and the physical structure of the solidified oil all influence the tendency of the oil to flow, the pour point of the oil is a guide to, and not an exact measure of, the temperature at which flow ceases under the service conditions of a specific system.

The pour point of wax-containing oils can be reduced by the use of special additives known as pour-point depressants which inhibit the growth of wax crystals. It is a recognized property of oils of this type that previous thermal history may affect the measured pour point ASTM Method D 97/IP 15 includes a section which permits some measurement of this thermal effect on waxy crystals.

The importance of the pour point, to the user of lubricants, is limited to applications where low temperatures are likely to influence oil flow. Obvious examples are refrigerator lubricants and automotive engine oils in cold climates. Any pump installed in outside locations where temperatures periodically fall below freezing should utilize lubricants with a pour point below some temperatures.

Flash and Fire Points

The flash-point test gives an indication of the presence of volatile components in an oil, and it is the temperature to which the oil must be heated under specified test conditions to give off sufficient vapor to form a mixture which will ignite in the presence of an open flame.

The fire point is the temperature to which the product must be heated under somewhat similar test conditions to cause the vapor/air mixture to burn continuously on ignition. The ASTM Test for Flash and Fire Points by Cleveland Open Cup (D 92/IP 36) can be used to determine both flash and fire points of lubricating oils, and it is the most generally used method for this purpose in the United States. In the United Kingdom, the ASTM Flash Point by Pensky Martens Closed Tester (D 93/IP 34) and open flash points (IP 35) are used widely.

Not only for the hazard of fire, but also as an indication of the volatility of the oil, the flash and fire points are significant in cases where high-temperature operations are encountered. In the case of used oils, the flash point is employed to indicate the extent of contamination by fuels or a more volatile oil. The flash point also can be used to assist in the identification of different types of base blends.

Relative Density (Specific Gravity) and API Gravity

Relative density and American Petroleum Institute (API) gravity are alternative but related means of expressing the weight of a measured volume of a product. Relative Density, ASTM Test for Density, Relative Density (Specific Gravity), or API Gravity of Crude Petroleum and Liquid Petroleum Products by Hydrometer Method (D 1298/IP 160), also known as specific gravity, is used widely outside the United States. In the United States, API gravity is used throughout the petroleum industry. The API gravity, ASTM Test for API Gravity of Crude Petroleum and Petroleum Products (Hydrometer Method) (D 287/IP 192), is based on a hydrometer scale which may be readily converted to the relative density basis by use of tables or formulas.

Both types of gravity measurements are used as manufacturing control tests. In conjunction with other tests, gravimetric measurements are used also for characterizing unknown oils, since they correlate approximately with hydrocarbon composition and, therefore, with the nature of the crude source of the oil.

Color

The color of a sample of lubricating oil is measured in a standardized glass container by comparing the color of the transmitted light with that transmitted by a series of numbered glass standards, Test for ASTM Color of Petroleum Products (ASTM Color Scale) (D 1500). The test is used for manufacturing control purposes and is important since the color is readily observed by the customer. The color of a lubricating oil is not always a reliable guide to product quality and should not be used indiscriminately by the consumer in writing specifications for purchases. Where the color range of a grade is known, a variation outside the established range indicates possible contamination with another product.

AUTOMOTIVE ENGINE OILS

Properties

The crankcase oil of automotive gasoline and diesel engines is used to lubricate and cool the pistons, cylinders, bearings, and valve train mechanisms. In some vehicles, the engine and gearbox or automatic transmission may be served by a common lubricant. Thus, the duty performed by an automotive engine oil is highly complex, and the oil needs to be formulated appropriately. As examples, the oil must contain sufficient oxidation inhibitors because even the best mineral oils react with oxygen at high temperatures to form sludge and varnish. Good detergent-dispersants are needed to suspend sludge and varnish forming material until they are removed by draining the oil. Water and combustion acids form during engine operation, and the corrosive wear and rusting which they cause must be counteracted with corrosion inhibitors which are usually designed to impart alkalinity to the oil. Many modern engines have high valve train loadings which require special antiwear additives in the lubricating oil. The compatibility of the various additives is also an important consideration.

Crankcase oil also has an impact on the control of exhaust emissions in today's ecology-conscious world. The operation of the positive crankcase ventilation valve, which prevents blowby gas venting to the atmosphere, is influenced by the quality of the oil used. Crankcase oils also can influence combustion chamber deposits, spark plug life, valve operation, engine wear, catalyst efficiency, and other factors influencing exhaust emissions.

To meet a particular ambient temperature condition, the viscosity of the engine oil is a main controlling property for manufacture and selection. Engine oils generally are recommended by vehicle manufacturers according to the SAE viscosity classification (J300). This classification sets the limits for the viscosity at 100°C (212°F) for all grades of oil and at −5°C to −35°C for W grades.

Multigrade engine oils for year-round service are sufficiently fluid at a low temperature to permit easy starting of the engine in winter conditions and still have an adequate viscosity at operating temperatures to provide for lubrication and to control oil consumption. The viscosity index (VI) of multigrade oils is typically in the range 120 to 160, while single grade oils are usually between 85 to 105.

The improved viscosity/temperature characteristics of multigrade oils enables, for example, an SAE 10W30 oil to be formulated to have characteristics of 10W oils at low temperature and 30 grade oils at higher temperatures. Multigrade oils containing viscosity index improver do not behave as Newtonian oils. The result is that the viscosity of multigrade oils generally is higher at −17.8°C (0°F) than is predicted by extrapolation from 100°C (212°F) and 40°C (104°F) viscosity values. The extent of the deviation varies with the type and amount of the viscosity index improver used. To better correlate with actual engine conditions, the viscosity of the SAE W grades is based on a measured viscosity using the Cold Cranking Simulator (ASTM D 2602) and the Mini-Rotary Viscometer (ASTM D 3829).

A number of factors are at work in an

engine to change the viscosity of the oil. Multigrade oils containing polymeric VI improver are subject to mechanical shearing resulting in a decrease in viscosity. The extent of this is dependent on the type of VI improver used since the base oil is relatively shear stable. The viscosity of automotive engine oils in service may also be decreased by fuel dilution and water but increased by oxidation and combustion products. Detergent/dispersant oils can keep these contaminants in suspension, but undesirable deposits (sludge, varnish, rust, etc.) can be formed on critical parts within the engine if oil drain periods are extended indiscriminately. Recommendations for oil-change intervals are usually made by the vehicle manufacturer, or, in the case of commercial vehicle fleets with known patterns of operation, the user may establish his own optimum change periods which are dependent on the quality level of the lubricating oil employed.

Engine Test Specifications and Procedures

Engine test methods used in the development of new formulations and for purchase specifications have originated from the following sources:

1. API/ASTM/SAE engine service classifications.
2. Institute of Petroleum tests (IP).
3. Coordinating European Council (CEC) tests.
4. Heavy duty diesel engine manufacturers tests.
5. Military specifications (U.S. Army and Navy, British Ministry of Defense, NATO, etc.).

Originally, the API defined the type of service for which an engine oil was designed, and the "MS" sequence tests were used to describe each type ("MS" denoted a service condition which was most severe for gasoline engines). In 1972, a joint API/ASTM/SAE system of nomenclature was adopted as a guide to the selection of engine oils for different service conditions.

Gasoline engine service conditions are designated by the letters SA, SB, SC, SD, SE, SF, and SG. Diesel engine oils are designated by CA, CB, CC, CD, CD-II, and CE. Currently, these original letter designations indicate increasing levels of service severity; however, as additional categories are adopted, this may not hold true. Performance criteria have been established for each designation using procedures which include published engine test Sequences IIA, IIB, IIC, IID, IIIA, IIIB, IIIC, IIID, IIIE, IV, V, VB, VC, V-D, VE, L-38, and 1H2 for gasoline engines. For diesel engines, tests are the L-4, L-38, L-1, LTD, IIA, IIB, IIC, IID, 1H, 1H2, 1D, 1G, 1G2, T-6, T-7, NTC-400, and 6V-53T. The designation and performance criteria are fully described in the SAE Standard Report J183, ASTM Research Report on engine oil performance classification (D2-1002), and ASTM Performance Specification for Automotive Engine Oils (D-4485).

The engine tests used to evaluate engine oil performance are included in *Multicylinder Test Sequences for Evaluating Automotive Engine Oils,* (ASTM STP 315), *Single Cylinder Engine Tests for Evaluating the Performance of Crankcase Lubricants* (ASTM STP 509), ASTM Research Report RR:D2:1194, and other ASTM Committee Section D.02.BO.02 reports. Many of the performance designations (oil categories) are obsolete in that they include performance tests where engine parts and/or test fuel and/or reference oils are no longer generally available, and the tests are no longer being monitored by the test developer or ASTM. Letter designations, where these situations are applicable, are SB, SC, SD, SE, SF, CA, and CB.

The energy-conserving potential of engine oils has also become important to the vehicle manufacturers. While not a part of the service condition designations, ASTM test methods (Five Car Test and Sequence VI) and performance categories have been established to designate energy-conserving engine oil performance levels. Most vehicle manufacturers recommend energy-conserving engine oils.

Important U.S. Army specifications include the MIL-L-46152 and MIL-L-2104 specifications, which are also used as performance references for commercial automotive engine oils. Vehicle manufacturers generally specify the most current performance criteria and specification engine oils during their warranty periods for cars and trucks.

Engine and bench tests for the evaluation of lubricants are also developed in Europe through the IP, CEC, and vehicle manufacturers. Some test procedures used are the Petter W-1, Ford Cortina, Fiat 132, DB OM 616, Bosch Injector, HTHSV, PSA TU-3, MWM-B, OM 352A, and the VW 1431. These lubricant tests along with some of the U.S. designed procedures are used to define performance targets G1, G2, G3 (gasoline engines), and D1, D2, D3, and PD1 (diesel engines).

Examination of Used Oils

Diesel fuel dilution, resulting from low-temperature or short-distance stop/start operation, can be estimated from measurements as determined by ASTM Test for Diesel Fuel in Used Lubricating Oils by Gas Chromatography (D 3524). Gasoline dilution can be measured by a distillation procedure, ASTM Test for Gasoline Dilution in Used Gasoline Engine Oils by Distillation (D 322/IP 23), or determined by ASTM Test for Fuel Dilution in Gasoline Engine Oils by Gas Chromatography (D 3525).

Low-temperature service conditions may also result in water vapor from combustion products condensing in the crankcase. This can be measured by distillation techniques, ASTM Test for Water in Petroleum Products and Bituminous Materials by Distillation (D 95/IP 74).

The extent and nature of the contamination of used automotive engine oil by oxidation and combustion products can be ascertained by determining the amounts of materials present in the lubricating oil which are insoluble in *n*-pentane and toluene, ASTM Test for Insolubles in Used Lubricating Oils (D 893). Both are expressed as percent by weight. Pentane insolubles include some oil-soluble resinous matter and all oil insoluble materials. Toluene insolubles include material from external contamination such as dirt, fuel carbon, and material from degradation of fuel, oil and additives, and engine wear and corrosion materials.

Where highly detergent/dispersant oils are under test, coagulated pentane insolubles and coagulated toluene insolubles may be determined by using methods similar to those just described but employing a coagulant to precipitate the very finely divided materials which may otherwise be kept in suspension by the detergent/dispersant additives.

Size discrimination of insoluble matter may be used to distinguish between finely dispersed, relatively harmless matter and the larger potentially harmful particles in oil (ASTM D 4055).

The metallic constituents (barium, calcium, magnesium, tin, silicon, zinc, aluminum, sodium potassium, etc.) of new and used lubricating oils can be determined by a comprehensive system of chemical analysis. For new lubricating oils, ASTM Test for Sulfated Ash from Lubricating Oils and Additives (D 874/IP 163) can be employed to check the concentration of metallic additives. These standard chemical procedures are time consuming to carry out, and—where the volume of samples justifies the purchase of such equipment—emission spectrographs, X-ray fluorescence spectrometers, atomic absorption, and other instruments which are much more rapid are available (see applicable ASTM/IP standards.)

The amount of reserve alkalinity remaining in the used oil can be determined by using ASTM Test for Base Number of Petroleum Products by Potentiometric Perchloric Acid Titration (D 2896/IP 276). Essentially, this is a titration method where, because of the nature of the used oil, an electrometric, instead of a color, end-point is used.

MARINE DIESEL ENGINE OILS

From the lubrication viewpoint, marine diesel engines are of two principal types. These are trunk-piston engines, in which the crankcase oil also lubricates the bearings and cylinders, and crosshead engines, in which the cylinders are separately lubricated. Marine diesel engines can also be classified as low-, medium-, and high-speed types with speeds of 0 to 50, 250 to 1000, and over 1000 rpm, respectively. The lower speed engines are less sensitive to fuel quality and can operate satisfactorily on residual fuels with high-sulfur contents, while the higher speed engines are generally more similar in design to automotive engines and use distillate fuels. Medium-speed engines vary in their fuel requirements according to the design.

The oil used in trunk-piston engines must have a viscosity suitable for lubricating the bearings and the cylinders. The trend in marine diesel engine design is towards smaller and lighter engines of higher specific output so that proportionally less space in a ship is required for the propulsion unit. Turbo-charging is used more frequently as a means of increasing the amount of power obtained from a given size of engine. This increases the heat input and oils with comparatively high levels of detergents, and antioxidants are required to maintain satisfactory engine cleanliness. The oxidation stability of trunk-piston engine oils can be measured by the single-cylinder CRC L-38 and the Petter W-1 tests. Detergency level is determined by one of the single-cylinder Caterpillar procedures.

Alkaline additives are required in the formulation of marine engine oils to neutralize potentially corrosive acids formed as a result of blow-by gases entering the crankcase. These additives must be capable of maintaining the alkalinity of the oil throughout the life of the charge. This is particularly important when high-sulfur (above 1 percent) residual fuels are used. Samples of oil drawn from the crankcase can be tested to assess the reserve of alkalinity remaining by determining the total base number of the oil. Other tests performed on the used engine oil, which serve as a guide to the suitability of the oil for further service, are the viscosity, flash point, pentane and benzene insolubles, and sulfated ash.

In the crosshead type engines, the crankcase oil lubricates the bearings and may also be used to cool the pistons. As well as having the appropriate viscosity, the oil also must have satisfactory oxidation resistance and good antifoam and anticorrosion properties. Small amounts of acidic contaminants entering the crankcase oil from the cylinders can be neutralized by using crankcase oils with a low level of alkalinity.

The cylinders of crosshead diesel engines are lubricated separately on an all-loss basis. The oil is injected into the cylinders through feed points around the cylinder and distributed by the scraping action of the piston rings. Excess oil collects in a scavenge space and runs off to the exterior of the engine. The oil used for this purpose is exposed to particularly high temperatures. As crosshead engines usually operate on residual fuels which frequently contain relatively high levels of sulfur, it is important that the cylinder oil has a sufficiently high level of alkalinity to neutralize acidic combustion products formed and thus to minimize the occurrence of corrosive wear. Some low-specific output engines are prone to exhaust-port blocking with carbonaceous combustion products and may also require special lubricating oils to reduce this tendency. Crosshead cylinder oils may be true solutions, in which the additives are dissolved in the oil, or they may contain finely divided additives suspended in the oil.

The deposit-forming tendencies due to thermal instability can be assessed by the Panel Coking Test when used in conjunction with other tests, and the available alkalinity of these oils can be measured.

The high-speed engines for fishing vessels, pleasure craft, or auxiliaries in large vessels require lubricating oils of a type similar to the higher detergency level automotive diesel engine lubricants. In some cases, automotive grades are recommended for use in these marine engines. However, since the sulfur contents of the fuels normally used in marine applications are higher than those of the fuels used for automotive purposes, it is important to ensure that the reserve of alkalinity is sufficient to neutralize the additional amounts of corrosive acids which may be formed.

INDUSTRIAL AND RAILWAY ENGINE OILS

A wide range of sizes and designs are employed for industrial engines, but the types of lubricants used and their related test procedures are similar to those described under the sections on automotive and marine engines. Crankcase systems for land-based engines may be larger in capacity than for marine purposes, and the oil charge, in these cases, would be expected to extend for longer periods of operation.

Engines with large cylinder bores (above 15.2 cm^3 or 6 in. diameter) usually have higher cylinder wall temperatures and require oils with high viscosities. Intermediate size engines generally use SAE 30 and 40 grade oils, while for the large engines SAE 50 oils may be recommended.

Because of space limitations locomotive engines have higher specific ratings than their industrial counterparts. The resultant higher bearing and cylinder temperatures, coupled with operating practices that include long idling periods followed by rapidly increased speed and load, make the lubricant performance criteria for locomotives relatively severe.

The importance of keeping railway locomotives in service for as long as possible before overhaul has led to the use of the regular checking of oil samples for the presence of wear metals. For this purpose direct reading spectrographs requiring a minimum of operator time, have been developed. A sudden increase in the amount of a particular metallic element present in the oil may indicate an incipient bearing failure.

New railway diesel oil formulations must pass specific tests of the railway diesel engine manufacturers and using railroads. Ultimate acceptance of an oil is based on extended satisfactory service performance.

GAS-TURBINE LUBRICANTS

Gas-turbine engines, originally developed for aviation use, are now being employed increasingly for industrial, marine, and automotive applications. The electricity generation industry, for example, uses a considerable number of gas turbines for standby and peak lopping purposes. Although more and more installations have begun to depend on this type engine for full time service, the requirement to conserve fuel resources may reverse this trend.

From the lubrication viewpoint, the most important features of gas turbines are the large volumes of high-temperature combustion gases which flow through the engine and the comparatively high-unit loads on the gearing as a result of the need to reduce weight to a minimum, particularly with turbines in aircraft. These features dictate the use of a lubricant of high-thermal and oxidative stability combined with good load-carrying properties. The latter property is particularly important where the engine oil also has to lubricate reduction gearboxes, such as those used for turbo-propeller or helicopter transmissions.

For aviation purposes, British and U.S. military specifications have had a strong influence in establishing quality criteria and performance levels for gas-turbine lubricants. The continuous development of engines for military and commercial use has resulted in increasingly high-bearing temperatures dictating the adoption of synthetic fluids rather than mineral oils.

Synthetic aircraft turbine lubricants were first used in the United States in 1952. The U.S. Air Force composed the first military performance specification, MIL-L-7808. Various changes in performance of qualified fluids have resulted in several revisions of this specification as indicated by a letter suffix, that is, MIL-L-7808 A, B, C, D, and F.

Pratt and Whitney engine tested most MIL-L-7808 type fluids in 1957 and subsequently formed their own qualified products list for commercial applications. General Electric Company (GM) and the Allison Division of General Motors Corporation also formed qualified products lists. The U.S. Navy and Pratt and Whitney initiated aircraft-turbine lubricant improvement programs in 1961. Their programs are generally referred to as Type "1½" and Type II, respectively. The military specification covering Type "1½" turbine lubricants is MIL-L-23699 (WEP), published by the U.S. Navy in early 1963. The Pratt and Whitney specification covering Type II aircraft-turbine lubricants is PWA 521-B Type II, published in mid-1963. The most significant difference between these two specifications is that MIL-L-23699 (WEP) requires qualified fluids to be shear-stable whereas the Pratt and Whitney specification does not. Otherwise, both specifications require almost identical performance.

The Type I synthetic lubricants were produced to meet the requirements of engines in the 1950s, but, with further advances in engine design together with the demand for increased periods between engine overhauls, the more thermally and oxidative stable Type "1½" and II lubricants were developed. Some of the physical and performance characteristics of the Type I,

"1 1/2," and II synthetic lubricants are shown in Table 1.

These specifications also include requirements for resistance to oxidation and corrosion. Other important properties of aviation gas-turbine lubricants are low volatility, foam resistance, seal compatibility, and hydrolytic stability.

Because of the safety aspect and the need to prolong periods between engine overhauls, the inservice condition of aviation lubricants is usually monitored by drawing regular samples for analysis. Metals analysis by emission spectroscopy can check whether or not any particular metal shows a sudden change in concentration in the used oil. This can indicate increased wear of a particular engine component and a possible need for replacement.

Industrial and marine gas-turbine installations vary in their severity of operating conditions relative to each other and aircraft service. They may run satisfactorily on high-quality, steam-turbine mineral oils or may need synthetic oils according to the particular service requirements. Automotive gas-turbine engines are being developed with the objective of using mineral oils, but, in most cases, high-bearing temperatures, particularly from heat soak along the drive shaft immediately after stopping the engine, have necessitated the use of either synthetic oils or mineral oil synthetic oil blends. In certain installations aboard ship or in industry, the very real danger of a fire from leakage of the lubricant may require the use of a fire-resistant fluid.

GAS ENGINE OILS

In areas where natural or liquefied petroleum gas (LPG) is available at a reasonable price level, these gases are finding increasing use as fuels for industrial engines. Gas engines range from large, relatively low output, low-temperature engines to small, high-speed, supercharged engines. Lubricant requirements vary with the engine design and operating conditions from uninhibited mineral oil, through mildly alkaline oxidation-inhibited detergent oils, to ashless highly detergent oils. Combustion in the large, low-output engines using natural gas or LPG fuels is relatively clean, and, since crankcases for these engines usually contain large quantities of oil, the operating conditions for the lubricant tend to be comparatively mild. For example, under certain conditions, an oil charge life of several years may be achieved.

The principal difference between the requirements of gas and other internal combustion engine oils is the necessity to withstand the degradation that can occur to the oil from accumulation of oxides of nitrogen which are formed by combustion. The condition of lubricating oils in large gas engines can be followed by measuring oil viscosity increase and determining changes in the neutralization number. Analytical techniques such as infrared spectroscopy and membrane filtration also can be used to check for nitration of the oil and buildup of suspended carbonaceous material.

The smaller gas engines generally op-

TABLE 1. Physical and performance characteristics of type I, 1 1/2, and II synthetic lubricants.

General Commercial Designation	Military Designation	Viscosity cSt at 99°C (210°F)	Limiting Low Temperature Viscosity	Load Carrying (Ryder Gear Test) lb./in. (kg/cm)	Use Temperature Bulk Oil
Type I	MIL-L-7808-F	3.0 (min)	13 000 cSt max at −53°C (−65°F)	1900 to 2200 (339 to 393)	to 149°C (300°F)
Type "1 1/2"	MIL-L-23699 (WEP) (USN)	5.0 (min)	13 000 cSt max at −40°C (−40°F)	2400 to 2700 (428 to 482)	to 204°C (400°F)
Type II	None[a]	5.0 (min)	13 000 cSt max at −40°C (−40°F)	1900 to 2400 (339 to 428)	to 204°C (400°F)

[a]Pratt and Whitney, PWA 521-B.

erate at high crankcase oil temperatures than occurs in the larger types, and lubricant degradation in these engines can be traced by viscosity, neutralization number, and insolubles determinations.

In many areas, the comparatively high-fuel cost for gas engines dictates that they operate at the maximum efficiency. In the case of two-cycle gas engines, relatively small amounts of port plugging (about 5 percent) can increase fuel costs to such an extent that engine overhaul becomes necessary. In these engines, port plugging can be caused by carbonaceous lubricant deposits and solid impurities in the combustion air. The lubricant should minimize port plugging from either of these causes and thus help to increase operation time between overhauls.

GEAR OILS

The range of uses for gear oils is extremely wide and includes industrial, automotive, marine, and aviation applications. Gears can vary from large, open types used in quarrying to very small instrument gears used for the control of aircraft. However, the primary requirement is that the lubricant provide satisfactory low wear and control or minimize other forms of damage such as pitting, scuffing, or rusting by maintaining a lubricant film between the moving surfaces.

Although gears are of many types including spur, helical, worm, bevel, hypoid, etc., they all function with some combination of rolling and sliding motion. The contact between the mating surfaces may be either along a line (as in the case of spur gear) or at a point (as for nonparallel, nonintersecting helical gear shafts). Although deformation of the metal will broaden the dimensions of the line or point contacts to areas of contact under service conditions, these areas are small in relation to the load on them. Therefore, the unit loadings of gear-tooth surfaces are relatively high compared with ordinary bearing surfaces. Gear teeth often transmit peak loads of 250 000 psi, 176×10^3 kg/cm^2, whereas sleeve bearing loads are usually in the range of 10 to 500 psi (0.07 to 35 kg/cm^2) and seldom exceed 3000 psi (210 kg/cm^2).

Where the gear loadings are comparatively light, straight mineral oils may be used as the lubricant, but, with increased unit loading, it is necessary to incorporate antiwear additives. For very highly loaded conditions in which shock is also experienced, special antiweld compounds are included in gear oil formulations.

It is necessary for a satisfactory balance to be maintained between the properties desired in the lubricant and the components used. For example, overactive chemical additives may promote undesirable wear by chemical attack. This type of wear is known as corrosive wear and results from the progressive removal of chemical compounds formed at the elevated temperatures on the tooth surface. Other types of gear wear are caused by fatigue, abrasion, and welding. In the case of metal fatigue, contact between the surfaces is not necessary for its occurrence. If the gear surfaces are subjected to stresses that are above the fatigue endurance limit of the metal, subsurface cracks can develop which may lead eventually to surface failure. Abrasion of the gear surface is caused either by the harder surface of the two cutting into the other or a hard contaminant or wear particle acting as the abrasive medium. Welding occurs when the severity of the load is high enough to cause complete breakdown of the lubricant film, and metal-to-metal contact occurs which results in transfer of metal between the surfaces. The correct choice of lubricant and satisfactory standards of cleanliness will minimize these wear effects.

Various methods are used to apply gear lubricants. Application may be by drip feed, splash, or spray. For large open gears, the lubricant must possess a sufficiently high viscosity and good adhesive properties to remain on the metal surface. This is particularly the case for applications such as strip coal mining and in the cement industry where gears may have to operate under wet and dirty conditions.

Large industrial gear sets using spur or helical gears operating under moderate loads are usually lubricated by circulating systems. The heavier viscosity turbine oils are generally suitable for this application. In enclosed systems, the temperatures reached may be high enough to necessitate the use of oils of good oxidation resistance. The oils should also possess satisfactory antifoam properties as well as good antitrust and demulsibility characteristics. Where higher loadings are encountered in industrial gears, lead, sulfur, and phosphorus compounds commonly are used to improve the load-carrying capacity of the lubricant.

For highly loaded spiral bevel, worm, or hypoid gears where sliding contact predominates over rolling contact between gear teeth, lubricating oils with special extreme pressure additives are used. Sulfur, chlorine, lead, and phosphorus compounds are used widely for this purpose.

The amount of active elements in new and used gear oils can be determined by "wet" analytical methods described previously, but there are also a number of instrumental techniques which enable the results to be obtained much more rapidly. Among these tests are polarographic, flame photometric, and X-ray fluorescence methods.

The analytical techniques described for measuring the elements associated with the load-carrying compounds present in gear oils do not identify the specific additives used; nevertheless, they are useful for controlling the quality of the finished products at blending plants. The same techniques can be used to determine similar additives and contaminants that may be present in used oils.

Since active sulfur is desirable for some extreme-pressure applications, ASTM Detection of Copper Corrosion from Petroleum Products by the Copper Strip Tarnish Test (D 130/IP 154) can indicate whether the formulation has a satisfactory level of sulfur activity. The copper strip test is used widely for the quality control of gear oils at blending plants.

Mechanical tests are used for assessing the extreme pressure, friction, and antiwear properties of gear oils in the laboratory. Examples of these are the Institute of Automotive Engineers (IAE) and Ryder gear rigs, the Timken Lubricant Tester, the SAE, David Brown and Caterpillar Disc (or Roller), 4-Ball, Falex, and Almen machines. The selection of the appropriate test machine depends on the application being considered. In spite of the variety of equipment available, the correlation of laboratory tests with practice is not precise, and, hence, the final evaluation of the lubricant needs to be made under controlled field conditions.

A number of industry and military specifications exist for automotive extreme pressure gear lubricants. Examples of these are the U.S. military MIL-L-2105 and 2105B specifications. The MIL-L-2105 specification is still used to indicate the performance level of oils, although the specification is now obsolete. The MIL-L-2105B specification describes an axle oil with a higher extreme pressure performance level, which is measured in the CRC L-42 Car Axle High-Speed Shock Test, and the CRC L-37 Truck Axle High-Torque Test. Also included in this specification are a Thermal and Oxidation Stability Test, the CRC L-33 Axle Moisture Corrosion Test, a Copper Corrosion Test, and a Channel Point Test to assess the low-temperature flow properties of the oil. The British Ministry of Defense Specification CS 3000B is based on the performance of a blended oil in full scale axle tests as well as a comprehensive series of laboratory bench tests including oxidation, foam, and storage stabilities. The axle tests specified are the High Torque Test (IP 232) and a modified High Speed Shock Test (IP 234).

In the case of farm tractor and highway construction equipment transmission lubricants, the fluid may be required to perform more than one function. For example, several tractor manufacturers recommend oils for use as both hydraulic and transmission fluids.

Several larger tractor models are now fitted with oil-immersed "wet" brakes. The oil used for this purpose may require special frictional characteristics to avoid the occurrence of severe vibration or "chatter" when braking is applied. Special additives are incorporated in oil formulations to meet this requirement.

Most major tractor manufacturers require approval of fluids for use in their tractors which is based on meeting physical and bench test specifications plus satisfactory performance in tractor tests conducted by the tractor manufacturer.

AUTOMATIC TRANSMISSION FLUIDS

The fluids used to lubricate automatic transmissions for passengers cars should facilitate the satisfactory operation of such components as the torque converter, planetary or differential gearing, wet clutches, servo-mechanisms, and control valves. The viscosity characteristics of the oil are extremely important. Since a minimum change of viscosity with temperature is desirable, viscosity index improvers are incorporated. Low viscosity improves the efficiency of the torque converter, but the lower limit is dictated by the viscosity required to protect the gearing. It is usual to employ fluids in the range of SAE 5W to 20W for automatic transmissions.

The shearing forces exerted by the automatic transmissions components, such as the pumps and clutches, tend to reduce the viscosity of the polymeric viscosity improvers incorporated in these fluids. It is important, therefore, that automatic transmission fluids have adequate shear-stability. To determine shear stability, full-scale road, dynamometer, or laboratory bench tests may be used with the viscosity compared before and after shearing. Because of the comparatively high temperatures reached in service, oils with very good oxidation and thermal stability are also necessary.

The two most important specifications for automatic transmission fluids in the United States have been the Ford M2C33E(F) and the GM "Dexron." The Dexron II and M2C33G specifications supercede the GM Dexron and Ford M2C33E(F) specifications, respectively. The new specifications place greater emphasis on oxidation stability, frictional durability, and antiwear and thermal stability characteristics of the fluid.

Because of the complex range of functions performed, automatic transmission fluids are highly complex products. Additives are used to enhance their performance. Load-carrying additives are used to protect the gearing and thrust washers. Smooth clutch engagements dictate the use of additives for foam control. Specific additives may be used to control the frictional properties of the lubricant since, with most automatic transmissions, static and dynamic friction properties are important for satisfactory operation of the clutches and brake bands. During selection of additives, a prime consideration is additive intercompatibility.

STEAM-TURBINE OILS

The primary purpose of lubricating oils for steam-turbine circulating systems is to provide satisfactory lubrication and cooling of the bearings and gears. It is also a common practice to design the turbine to have the lubricating oil also function as a governor-hydraulic fluid. The viscosity of the oil is important for both of these functions, but the loading of the gear is the major factor in the choice of lubricant. A sufficiently thick film of oil must be maintained between the load-bearing surfaces; the higher the load on the gears, the higher the viscosity required. However, as the circulated oil also acts as a coolant for the bearings, it is necessary to have as low a viscosity as possible consistent with the lubrication requirements.

Since steam-turbine oils generally are required to function at elevated temperatures (for example, 71.1°C or 160°F), it is most important that the oxidation stability of the oil is satisfactory, otherwise, the service life of the oil will be unduly short, ASTM Test for Oxidation Characteristics of Inhibited Mineral Oils (D 943). Oxidation inhibitors are added to the base oil to improve this characteristic. Lack of oxidation stability results in the development of acidic products which can lead to corrosion (particularly of bearing metals) and also affect the ability of the oil to separate from water. Oxidation also causes an increase in viscosity and the formation of sludges which restricts oilways, impairs circulation of the oil, and interferes with the function of governors and oil relays. Correctly formulated turbine oils have excellent resistance to oxidation and will function satisfactorily for years without changing the system charge.

Turbine-oil systems usually contain some free water as a result of steam leaking through glands and then condensing. Marine systems may also have salt water present due to leakage from coolers. Because of this, rust inhibitors are almost always incorporated in the formulation. Rust preventing properties may be measured by ASTM Test for Rust-Preventing Characteristics of Inhibited Mineral Oil in the Presence of Water (D 665/IP 135).

The presence of water in turbine systems tends to lead to the formation of emulsions and sludges containing water, oil, oil oxidation products, rust particles, and other solid contaminants which can seriously impair lubrication. The lubricating oil, therefore, should have the ability to separate from water readily and resist emulsification. Approximate guides to the water-separating characteristics can be gained by ASTM Test for Water Solubility of Petroleum Oils and Synthetic Fluids (D 1401/IP 79).

Although systems should be designed to avoid entrainment of air in the oil, it is not always possible to prevent. The formation of a stable foam increases the surface area of the oil which is exposed to small bubbles of air thus assisting oxidation. The foam can also cause loss of oil from the system by overflow. Defoamants are usually incorporated in turbine oils to decrease their foaming tendency. Foaming tendency can be measured by ASTM Test for Foaming Characteristics of Lubricating Oils (D 892/IP 146). Air release is also an important property if a soft or spongy governor system is to be avoided. A careful choice of type and amount of defoamant will provide the correct balance of foam protection and air release properties.

Marine turbine gearing design has advanced to the stage where increased loading has permitted a decreased size of the gear train. This saving in space is very desirable, particularly in naval vessels, but turbine oils for such applications may also require a moderate level of extreme-pressure properties. Load-carrying properties for turbine oils may be measured by ASTM Test for Load-Carrying Capacity of Petroleum Oil and Synthetic Fluid Gear Lubricants (D 1947), which uses a Ryder gear rig, or by IP 166 which utilizes an IAE gear rig.

HYDRAULIC OILS

The operation of many types of industrial machines can be controlled conveniently by means of hydraulic systems which consist essentially of an oil reservoir, a pump, control valves, piping, an actuator, and sometimes an accumulator. The wide range of hydraulic applications encountered necessitates the use of a variety of pump designs including gear, vane, axial, and radial piston types which, in turn, utilize various metallurgical combinations. The hydraulic fluid is required to transmit pressure and energy, minimize friction and wear in pumps, valves and cylinders, minimize leakage between moving components, and protect the metal surfaces against corrosion.

To obtain optimum efficiency of machine operation and control, the viscosity of the oil should be low enough to minimize frictional and pressure losses in piping. However, it also is necessary to have a sufficiently high viscosity to provide satisfactory wear protection and minimize leakage of the fluid. High-viscosity index fluids help to maintain a satisfactory viscosity over a wide temperature range. The antiwear properties of high-quality hydraulic oils usually are improved by the incorporation of suitable additives in the formulation.

Since the clearances in pumps and valves tend to be critical, it is important to provide adequate filtration equipment (full flow or bypass or both) to maintain the system in as clean a condition as possible and thus minimize wear. The oil should have good oxidation stability to avoid the formation of insoluble gums or sludges; it should have good water separation properties, and, because air may be entrained in the system, the oil should have good air-release properties and resistance to foaming. Similarly, good rust protection properties will assist in keeping the oil in a satisfactory condition.

While petroleum oils, properly formulated, are excellent hydraulic fluids, they are flammable, and their use should be avoided in applications where serious fires could result from oil leakage contacting an open flame or other source of ignition. Fire-resistant fluids should be used in such applications. The major fire resistant fluids are (1) the synthetics (phosphate esters, silicones, and silicates), (2) the synthetics-petroleum oil blends (to reduce cost), (3) the water-glycol solutions, and (4) the petroleum oil invert emulsions (water-in-oil).

Over the years, a number of tests have been used to evaluate the fire-resistant properties of such fluids under a variety of conditions. Most tests involve dripping, spraying, or pouring the liquid into a flame or on a hot surface of molten metal. ASTM recently has developed and published a method in which the fluid is sprayed as a mist into the flame of a laboratory burner, ASTM Test for Mist Spray Flammability of Hydraulic Fluids (D 3119).

Fire-resistant fluids, or safety devices on hydraulic systems, are used widely in the coal mining industry. The use of such fluids also is expanding in the metal cutting and forming, lumber, steel, aluminum, and aircraft industries.

OTHER LUBRICATING OILS

While many of the properties which have been discussed earlier apply to other types of lubricating oils, there are a number of industrial applications for which special performance requirements are demanded. The following are examples.

Air Compressor Oils

In addition to possessing the correct viscosity for satisfactory bearing and cylinder lubrication, very good oxidation resistance is required to avoid degradation of the lubricant in the presence of heated air. This is particularly important where discharge temperatures are high, since carbon and oxidized oil deposits may autoignite if exposed continuously to temperatures above 148°C (300°F). The fire potential that exists under these conditions make low volatility and high-autoignition values equally or more important than high-flash or fire points. As an added safeguard against compressor fires, a fire resistant fluid (phosphate ester) should be considered.

In air compressor lubrication, condensed water is present frequently. For this reason, the oil must possess properties that ensure that the oil rather than water wets the metal surfaces. Also, to avoid the accumulation of invert water-in-oil emulsions in the after coolers, the water should separate out rather than form an emulsion.

Refrigerator Compressor Oils

The efficiency of compression-refrigeration systems can be influenced directly by the properties of the lubricant. This is because the oil used for cylinder lubrication tends to be carried over into the system where it can have a detrimental effect on the efficiency of the evaporator and associated equipment. The lubricating oil, therefore, must effectively minimize friction and wear in the compressor and prevent the formation of undesirable deposits. In addition, the lubricant should have no adverse effects on the operation of the condenser expansion valve, or evaporator. Low-temperature viscosity must be balanced against the need to protect the cylinder against wear. Adequate oxidation resistance and thermal stability to resist the high temperatures encountered at the compressor discharge are also necessary. In ammonia systems, it is important that the pour point of the oil is below the evaporator temperatures to avoid the congealing of the oil on evaporator heat-transfer surfaces. With Freon systems, a low-Freon floc point of the oil indicates freedom from the likelihood of waxy deposits which could otherwise interfere with the satisfactory operation of the expansion valve and lower the rate of heat transfer.

Steam Cylinder Oils

Steam engines, pumps, forging hammers, and pile drivers are among the equipment using steam cylinders. The performance of this type of machinery is affected directly by the efficiency of the lubrication of the valves, piston rings, cylinder walls, and rods. The selection of the lubricating oil is influenced by the steam temperatures encountered, the moisture content of the steam, the cleanliness of the steam (that is, possible contamination by solids), and the necessity for the oil to separate from the exhaust steam or condensate. Excessive oxidation of the oil in service could cause a buildup of deposits in the stem and rod packings and result in shutdown of the equipment for cleaning.

Machine Tool Tableway Lubricating Oils

Satisfactory lubrication of the ways and slides of machine tools is important in maintaining the precision of equipment designed to work to close tolerances. The movement of worktables, workheads, tool holders, and carriages should be facilitated by the lubricant so that the control is smooth and precise. The characteristics required for these oils include a suitable viscosity to enable ready distribution of the oil to the sliding surface, while ensuring that the necessary oil films are formed at traverse speeds under high-load conditions. Static friction must be minimized, and the oil should prevent the alternate sticking and slipping of moving parts, particularly at very low speeds. Especially when the position of these surfaces is in the vertical plane, good adhesive properties are required to maintain an adequate film on intermittently lubricated surfaces.

Applicable ASTM/IP Standards

ASTM	*IP*	*Title*
	19	Demulsification Number—Lubricating Oil
	35	Flash Point (Open) and Fire Point by Means of the Pensky-Martens Closed Tester
D 92	36	Flash and Fire Points by Cleveland Open Cup
D 93	34	Flash Point by Pensky-Martens Closed Tester
D 95	74	Water in Petroleum Products and Bituminous Materials by Distillation
D 97	15	Pour Point of Petroleum Oils
	110	Barium in Lubricating Oil
	111	Calcium in Lubricating Oil
	114	Oxidation Test for Turbine Oils
	117	Zinc in Lubricating Oil
	120	Lead, Copper and Zinc in Lubricating Oils
D 129	61	Sulfur in Petroleum Products (General Bomb Method)
	148	Phosphorus in Lubricating Oil, Additives, and Concentrates
D 130	154	Detection of Copper Corrosion from Petroleum Products by the Copper Strip Tarnish Test
	166	Load-Carrying Capacity Test for Oils—IAE Gear Machine
	175	Engine Cleanliness—Petter W1 Spark-Ignition Test
	176	Oil Oxidation and Bearing Corrosion—Petter W1 Spark-Ignition
D 322	23	Gasoline Diluent in Used Gasoline Engine Oils by Distillation
D 341		Viscosity-Temperature Charts for Liquid Petroleum Products
D 445	71	Kinematic Viscosity of Transparent and Opaque Liquids (and the Calculation of Dynamic Viscosity)
D 665	135	Rust-Preventing Characteristics of Inhibited Mineral Oil in the Presence of Water
D 808		Chlorine in New and Used Petroleum Products (Bomb Method)
D 811		Chemical Analysis for Metals in New and Used Lubricating Oils
D 874	163	Sulfated Ash from Lubricating Oils and Additives
D 892	146	Foaming Characteristics of Lubricating Oils
D 893		Insolubles in Used Lubricating Oils
D 943	157	Oxidation Characteristics of Inhibited Mineral Oils

Applicable ASTM/IP Standards *(continued)*

ASTM	*IP*	*Title*
D 974	139	Neutralization Number by Color-Indicator Titration
D 1091		Phosphorus in Lubricating Oils and Additives
D 1317	118	Chlorine in New and Used Lubricants
D 1401		Water Solubility of Petroleum Oils and Synthetic Fluids
D 1500		ASTM Color of Petroleum Products (ASTM Color Scale)
D 1947		Load-Carrying Capacity of Petroleum Oil and Fluid Gear Lubricants (Erdco, Ryder, WAAO Machines)
D 2270	226	Calculating Viscosity Index from Kinematic Viscosity at 40 and 100°C
D 2271		Preliminary Examination of Hydraulic Fluids (Wear Test)
D 2272		Continuity of Steam-Turbine Oil Oxidation Stability by Rotating Bomb
D 2422		Recommended Practice for Viscosity System for Industrial Fluid Lubricants
D 2500	219	Cloud Point of Petroleum Oils
D 2602		Apparent Viscosity of Engine Oils at Low Temperature Using the Cold-Cranking Simulator
D 2619		Hydrolytic Stability of Hydraulic Fluids (Beverage Bottle Test Method)
D 2670		Measuring Wear Properties of Fluid Lubricants (Falex Method)
D 2711		Demulsibility Characteristics of Lubricating Oils
D 2782		Measurement of Extreme-Pressure Properties of Lubricating Fluids (Timken Method)
D 2783		Measurement of Extreme-Pressure Properties of Lubricating Fluids (Four-Ball Method)
D 2882		Method for Indicating the Wear Characteristics of Petroleum and Non-Petroleum Hydraulic Fluids in a Constant Vane Pump
D 2893		Oxidation Characteristics of Extreme-Pressure Oils
D 2896	276	Total Base Number of Petroleum Products by Potentiometric Perchloric Acid Titration
D 2982		Detecting Glycol-Base Antifreeze in Used Lubricating Oils
D 2983		Low-Temperature Viscosity of Automotive Fluid Lubricants Measured by Broofield Viscometer
D 3119		Mist Spray Flammability of Hydraulic Fluids

9

Lubricating Greases

INTRODUCTION

THE PRIMARY FUNCTION OF LUBRICANTS is to increase the efficiency of machinery by reducing friction and wear. Secondary functions include the dissipation of heat and removal of contaminants. Fluid lubricants are difficult to retain at the point of application and must be replenished frequently. If, however, a fluid lubricant is thickened, its retention is improved. A lubricating grease is simply a lubricating fluid which has been gelled by means of a thickening agent so that it may be retained more readily in the required area.

Lubricating greases have a number of advantages over lubricating fluids. Some of these are: (1) dripping and spattering are nearly eliminated; (2) less frequent applications are required; (3) greases are easier to handle; (4) less expensive seals are needed; (5) greases form a seal in many cases and keep out contaminants; (6) they adhere better to surfaces; (7) they reduce noise and vibration; (8) some grease remains even when relubrication is neglected; and (9) greases are economical.

Many ways have been devised to thicken lubricating oils. Soaps were the first thickeners used and still have wide application. Other thickeners that have been employed include polymers, clays, silica gel, etc.

The lubricating fluids which have been thickened to form greases vary widely. By far the largest volume of greases in use today are those made with mineral oils thickened with soaps. Many types of mineral oils are used, for example, naphthenic, paraffinic, blended, hydrocracked, hydrogenated, solvent refined, highly refined, etc. In addition to mineral oils, many other fluids, such as esters, diesters, silicones, polyethers, and synthetic hydrocarbons, can be used.

When selecting a grease, one should first determine the type and viscosity of the lubricating fluid required. The fluid provides the lubrication; however, grease properties are based dominantly on the characteristics of the thickener system. Each thickener system has its own unique characteristics. Properties of the various soap-type greases will be discussed in later paragraphs.

COMPOSITION

Grease has been defined as a gelled lubricating fluid. The basic steps which are required to make a grease include dissolving the thickener system in the fluid, thoroughly mixing, cooling the mixture to permit proper crystallization of the thickener, blending in the desired additives and inhibitors, and then final finishing and processing. The inhibitors enhance the performance of the grease. Included in the finishing and process steps are homogenization or deaerating or both where needed.

The fatty materials used for soap formation may be of animal or vegetable origin. The type of fatty materials used affects the properties of the soap and the grease. Improved or special properties or both may be obtained by the use of mixtures or blends of fatty materials.

Soaps are formed by the simple saponification of an acid and base to form a salt and water. If the acid is a fatty acid, its salt is called a soap. If the metallic moiety of the base is monovalent a simple soap is made, for example, lithium stearate. If the metal is polyvalent, a complex soap having unique properties is formed. For example, if two dissimilar acids are attached to the same metallic element, calcium stearate acetate can be formed. Mixed base greases consist of a mixture of two different thickener systems.

Salts of fatty acids (soaps) are present in greases in the form of fibers. The structure and size of these fibers, that is thickness and length, depends upon the metallic moiety and the conditions under which they are formed. In general, fibers may vary from about 1 to 100 μm in length with a length/diameter ratio of about 10 to 100 μm. Large, coarse fibers do not absorb fluids as well as fine, closely knit fibers. Thus, higher percentages of coarse fibered soaps are required to make greases having the same consistency as those made with fine fibered soaps.

PROPERTIES

Calcium Soap Greases

The earliest known greases were made with calcium soaps. Calcium soaps are water resistant, mechanically stable, and inexpensive. The greatest shortcoming of calcium soap greases is the low-melting points (dropping points) which are usually about 95°C (200°F).

Sodium Soap Greases

Sodium soap greases have higher dropping points [about 205°C (400°F)] than calcium greases. However, they are not water resistant and emulsify in the presence of water. They are not normally compatible with other greases. Sodium soap greases do have an inherent ability to provide rust protection.

Lithium Soap Greases

Lithium greases were the first so-called multipurpose greases that offered both the water resistance of calcium soap greases and the high-temperature properties of sodium soap greases. Lithium soap greases have dropping points of about 175°C (350°F). They also have good mechanical stability; that is, they soften very little upon working.

Complex Soap Greases

Complex soap greases are noted for their high-dropping points [260°C (500°F) and higher]. They have good resistance to water and usually have good mechanical stability properties.

Clay Greases

Clay thickened greases have been referred to as nonmelting greases since they tend to decompose [at temperatures above 290°C (550°F)] before reaching their dropping point. They usually have poorer mechanical stability and water resistant properties than soap greases. They also are more difficult to inhibit with extreme pressure additives.

EVALUATION OF PROPERTIES AND SIGNIFICANCE

A number of tests have been developed and standardized by American Society for Testing and Materials (ASTM) and the Institute of Petroleum (IP) which describe the properties of performance characteristics of lubricating greases. Since these tests are conducted in laboratories under well-defined conditions, they are used primarily as screening tests. Some of the grease tests do give an indication of what a grease might do in service, but direct correlation between laboratory and field performance is rarely possible, since the tests never exactly duplicate service conditions.

Consistency

Consistency has been defined as the degree to which a plastic material, such as a lubricating grease, resists deformation under the application of force. The standard method for measuring grease consistency is the penetration test. Consistency is reported in terms of ASTM cone penetration, National Lubricating Grease Institute (NLGI) number, or apparent viscosity. Cone penetrations and NLGI number are discussed in the following paragraphs. Apparent viscosity is included in the section on Shear Stability.

Cone Penetration

The standard method for the determination of the penetration of a normal grease sample is ASTM Test for Cone Penetration of Lubricating Grease (D 217/IP 50). In this method, a double-tapered cone of prescribed geometry sinks under its own weight into a sample of greases at 25°C (77°F) for 5 s. The depth of penetration is measured in tenths of a millimeter, and it is the penetration value. Stiff greases will have very low-penetration values, whereas a soft grease will have a high-penetration number.

The penetration number for small samples of grease may be determined by using ASTM Test for Cone Penetration of Lubrication Grease Using One-Quarter and One-Half Scale Cone Equipment (D 1403/IP 310). An equation is used to convert the penetrations obtained in this test to equivalent penetrations for the full scale test.

The following paragraphs describe four procedures used to obtain penetrations.

Undisturbed Penetration

The penetration is measured in the container as received without any disturbance. This value is significant in determining the extent to which a grease may harden or soften in the container in shipment or storage or both.

Unworked Penetration

This value is obtained where the grease is transferred from the container, with only a minimum amount of disturbance, to the cup of the grease worker. This result is not always reliable, since the amount of disturbance cannot be controlled nor repeated exactly. It may be significant to indicate consistency variances in transferring a grease from the container to the equipment.

Worked Penetration

Worked penetration is the standard penetration value for a grease. It is measured after a grease has been worked for 60 double strokes in the grease worker. This method is more reliable, since disturbing the grease is more nearly standardized by the working process.

Prolonged Worked Penetration

This value is obtained after a sample has been worked for a prolonged period in the grease worker, that is, 10 000, 50 000, 100 000, etc. double strokes. After prolonged working in the grease worker, the sample and worker are brought back to penetration test temperature 25°C (77°F) in 1.5 h. It is then worked for 60 double strokes and the penetration is measured. This test is significant, since it can indicate the degree of mechanical or shear stability of a grease. It will be described later.

NLGI Grades

On the basis of worked penetrations, the NLGI has standardized a numerical scale as a means of classifying greases in accordance with their consistency. This scale is shown below in order of increasing hardness.

ASTM Worked Penetrations	NLGI Grade
445 to 475	000
400 to 430	00
355 to 385	0
310 to 340	1
265 to 295	2
220 to 250	3
175 to 205	4
130 to 160	5
85 to 115	6

Shear Stability

The ability of a grease to resist changes in consistency during mechanical working is referred to as shear or mechanical stability. Two methods have been standardized to evaluate the stability of a grease to working.

Prolonged Worked Penetration

ASTM D 217/IP 50, described previously, is used before and after prolonged working in a grease worker to determine the change in grease consistency.

Roll Stability

Roll stability is determined by ASTM Test for Roll Stability of Lubricating Grease

(D 1831) in conjunction with ASTM D 1403/IP 310.

After a worked penetration has been measured on a grease sample by ASTM D 1403, 50 g of the worked grease are placed into a horizontally-mounted cylinder containing an 11-lb steel roller. The cylinder is rotated at 165 rpm for 2 h. The inner roller rolls over the grease, working it during the test. After the test, the penetration of the grease is once again measured by D 1403, and the difference between the before and after penetration is noted.

In both of these tests, the change in consistency is reported as either the absolute change in penetration values or the percent change as outlined in ASTM D 1831.

Roll stability tests are significant because they may show a directional change that may occur in the consistency of a grease during service. No accurate correlations have been determined.

Apparent Viscosity

Grease is by nature a nonNewtonian material. It is characterized by the fact that flow is not initiated until stress is applied. Increases in shear stress or pressure produce disproportionate increases in flow. The term "apparent viscosity" is used to describe the observed viscosity of these materials and is measured in poises. Since this observed viscosity varies with both temperature and shear rate, it must be reported at a specific temperature and shear rate. Apparent viscosity is determined by ASTM Test for Apparent Viscosity of Lubricating Greases (D 1092).

In this test, a sample of grease is forced through a series of eight capillary tubes by a floating piston actuated by a hydraulic system using a two-speed gear pump. Poseutille's equation is used to calculate the apparent viscosity. The results are expressed as a logarithmic graph of apparent viscosity versus shear rate at a constant temperature, or apparent viscosity versus temperature at a constant shear rate. This equipment also has been used to measure the pumpability of greases at low temperature.

The apparent viscosity test is significant because it provides an indication of ease of handling and dispensing at a specified temperature. It also is used as an indication of the directional value of starting and running torques of grease lubricated mechanisms. Specifications may include limiting values of apparent viscosity for greases to be used at low temperature.

Dropping Point

The dropping point of a grease is the temperature at which it passes from a semisolid to a liquid.

Two similar procedures are used to determine the dropping point of grease. In both methods, a prescribed film of grease is coated on the inner surface of a small cup whose sides slope toward a hole in the bottom. With ASTM Test for Dropping Point of Lubricating Grease (D 566/IP 132), the sample is heated at a prescribed rate until a liquid drop falls from the cup. In ASTM Test for Dropping Point of Lubricating Grease Over Wide Temperature Range (D 2265), the sample is introduced into a preheated environment so that the heating rate is controlled more uniformly. In both tests, the difference in temperature between the grease in the cup and the environment are taken into account in calculating the dropping point of the grease. Some greases containing thickeners other than soaps may not separate oil nor melt.

The dropping point is useful (1) in identifying the type of thickener used in a grease, (2) as an indication of the maximum temperature to which a grease can be exposed without complete liquefaction or excessive oil separation, and (3) in establishing bench marks for quality control.

Although greases normally do not perform satisfactorily at temperatures above the dropping point, other factors are involved. High-temperature performance can depend on the application method and frequency and whether or not a softened grease is retained at the point of application by proper seals. Whether the high temperature is continuous or intermittent, etc. also depends upon the stability and evaporation properties of the grease.

Dropping point is useful, but it has no direct bearing on service performance unless such a correlation has been established.

Oxidation Stability

ASTM Test for Oxidation Stability of Lubricating Greases by the Oxygen Bomb Method (D 942/IP 142) was designed to predict shelf storage life of greases in prepacked bearings. In this test, 5 glass dishes are filled with 4 g of grease, each for a total of 20 g. These dishes are then sealed in a bomb on a rack, and the bomb is pressurized to 7.7 kPa/cm^2 (110 psi) with oxygen. The bomb is heated in a bath at 99°C (210°F) to accelerate oxidation. The amount of oxygen absorbed by the grease is recorded in terms of pressure drop over a period of 100 h and, in some cases, 500 h. The pressure drop is a net result of absorbed oxygen and any gases or by-products released from the grease.

Care must be exercised in the interpretation of data derived from the oxidation bomb test. Additives incorporated into the grease can produce misleading results because they may also react with oxygen. As an example, sodium nitrite is sometimes added to grease to serve as a rust inhibitor. In the oxidation bomb test, this material reacts with oxygen to form sodium nitrate. In this instance, the drop in pressure is not indicative of the amount of oxidation of the grease alone.

The oxidation bomb test is significant because it provides an indication of the shelf life of a thin film of grease, as on machine parts or prelubricated bearings, stored for long periods of time. It is a static test and not intended for predicting the performance of grease under dynamic conditions. Nor is this test intended for predicting the stability of grease stored in sealed commercial packages.

Effect of Copper or Oxidation Rate

ASTM Test for Effect of Copper on Oxidation Stability of Lubricating Greases by the Oxygen Bomb Method (Intent to Withdraw) (D 1402) is used to determine the effect of copper on oxidation rate. The procedure is the same as the oxidation bomb test (ASTM D 942/IP 142) discussed in the preceding paragraphs, except that clean, freshly prepared copper strips are placed on edge in each of the five dishes of grease. The test is discontinued when the pressure has dropped to 3.85 kPa/cm^2 (55 psi) or at some predetermined time period.

The results from this procedure indicate the catalytic effect of copper and its alloys on the acceleration of grease oxidation under static conditions for long periods of time. Such conditions occur when thin coatings of grease on metal parts or bearings, in the presence of copper or copper alloys, are stored under shelf conditions for extended time periods.

The method should not be used to predict the stability of grease on contact with copper under dynamic conditions or when stored in sealed commercial packages.

Oil Separation

Greases differ markedly in their tendency to liberate oil. Although opinions differ on whether or not lubrication is dependent on oil bleeding, excessive liberation of free oil during storage is to be avoided.

The fluid can be squeezed out of a grease at varying rates depending on the gel structure, the nature and viscosity of the lubricating fluid, and the applied pressure and temperature.

ASTM Test for Oil Separation from Lubricating Greases During Storage (D 1742) is used to determine the tendency of lubricating greases to separate oil during storage at 25°C (77°F). It is not suitable for use with greases softer than NLGI No. 1 consistency because of a tendency for the grease to seep through the screen.

The test is useful because the results correlate with oil separation which occurs in containers of grease during storage under corresponding ambient temperatures. It should not be used to predict the oil separation of grease under dynamic service conditions.

Evaporation Loss

Exposure of a grease to high temperatures may cause evaporation of some of the liquid lubricant causing the remaining grease to become drier and stiffer or leading to other undesirable changes in the grease structure. Greases containing some light viscosity oils for good low-temperature performance may show evaporation losses at the higher temperatures. Evaporation also may cause problems where vapors may be hazardous or combustible or interfere with operations as in gas-cooled reactors.

Evaporation loss is determined by ASTM Test for Evaporation Loss of Lubricating Greases over Wide-Temperature Range (D 2595). The test is useful for the determination of loss of volatile materials from a grease at temperatures between 93 and 316°C (200 and 600°F). It also is used to compare the evaporation losses of greases intended for similar service. (The grease which shows the least loss to evaporation, all other factors being equal, will probably perform longer in service.)

Results of ASTM D 2595 may not be representative of volatilization which may occur in service.

Rust Prevention

Greases must not be corrosive to metals they contact and should not develop corrosion tendencies with aging or oxidation. A method for assessing rust prevention by greases is ASTM Test for Corrosion Preventive Properties of Lubricating Greases (D 1743).

In the ASTM D 1743 method, a tapered roller bearing is packed with grease and, following a short run-in period, dipped into distilled water and stored above the water in 100 percent relative humidity at 52°C (125°F) for 48 h. The bearing is then cleaned and examined for corrosion.

The significance of this test is that it indicates those greases capable of preventing rust and corrosion in static or storage condition. The correlation with service conditions, particularly under static conditions, is considered to be quite good.

Lead in Greases

ASTM Test for Lead in New and Used Greases (D 1262) is used to determine the lead content of new and used greases containing 0.1 percent or more lead. It is also applicable for the determination of lead in fractions separated from a grease by means of appropriate solvents. Other metallic elements—sulfur, chlorine, phosphorus—in amounts commonly found in greases do not interfere in this method.

Lead content may be indicative of the amount of soaps or lead containing additives in a grease. ASTM D 1262 can be used to monitor the level of these components.

Water Washout

The ability of a grease to resist washout under conditions where water may splash or impinge directly on a bearing is an important property in the maintenance of a satisfactory lubricating film. ASTM Test for Water Washout Characteristics of Lubricating Greases (D 1264) evaluates the resistance of a lubricating grease to washout by water from a bearing at 38°C (100°F) or 79°C (175°F).

This test method uses a 204 K Conrad type 8-ball bearing equipped with front and rear shields that have a specified clearance. It is packed with 4 g of the test grease and then rotated at 600 rpm for 1 h while a jet of water at either 38°C (100°F) or 79°C (175°F) impinges on the bearing housing. The bearing is dried, and the percent loss by weight of grease is determined.

The test serves only as a measure of the resistance of a grease to water washout. It should not be considered the equivalent of a service evaluation unless such correlation has been established. Even comparative results between different greases may not predict the relative performance of the two greases in actual field use.

Extreme Pressure Timken Method

The ASTM Test for Measurement of Extreme Pressure Properties of Lubricating Grease (Timken Method) (ASTM D 2509) may be used to determine the load carrying capacity of a grease.

In this test, a tapered roller bearing cup is rotated against a stationary hardened steel block. Fixed weights force the block into line contact with the rotating cup through a lever system with a mechanical advantage of ten. The "OK Value" is the maximum load the lubricant film will withstand without rupturing and causing scoring in the contact zone after a 10-min run.

The test is a rapid method which may be used to differentiate between grease having low, medium, or high levels of extreme pressure properties. The results of this test do not necessarily correlate with results from field service.

Extreme Pressure Four-Ball Test

ASTM Measurement of Extreme Pressure Properties of Lubricating Grease (Four-Ball Method) (D 2596) is also used to determine the load-carrying properties of lubricating greases. With this procedure two evaluations may be made: (1) the Load-Wear Index (formerly called Heitz-Mean Load), and (2) the Weld Point.

The test was developed to evaluate the extreme pressure, antiwear, and antiweld properties of a lubricant. The tester is operated with one steel ball under load rotating against three steel balls held stationary to form a cradle. The grease under test covers the area of contact of the four balls. Unit pressures as high as 6 kPa (6.89×10^6 kPa/cm^2) (1 000 000 lb/in.^{2}) can be attained.

The procedure involves the running of a series of 10-s tests over a range of increasing loads until welding occurs. During a test, scars are formed in the surfaces of the three stationary balls. The diameter of the scar depends upon the load, speed, test duration, and lubricant. The scars are measured under a microscope having a calibrated grid. From the scar measurements the Load-Wear Index is calculated.

The significance of this test is that it is a rapid method which may be used to differentiate between greases having low, medium, or high levels of extreme pressure properties.

Lubricating greases with a fluid component that contains a silicone or a halogenated silicone oil can not be evaluated with D 2596. In addition, the results of this test do not necessarily correlate with results from field service.

Wear Preventive Characteristics of Grease

The determination of wear preventive characteristics of greases in sliding steel-on-steel applications is made with ASTM Test for Wear Preventive Characteristics of Lubricating Grease (Four-Ball Method) (D 2266). With this test a steel ball is rotated under load against three stationary steel balls with grease lubricated surfaces. The diameters of the wear scars that occur on the stationary balls are measured after completion of the test.

The test is significant because it can be used to determine the relative wear-preventing properties under the test conditions.

ASTM Test D 2266 is limited in the following respects.

1. If test conditions are changed, the relative ratings may change.

2. Wear characteristics are not predicted for metal combinations other than steel.

3. No differentiation can be made between extreme pressure and nonextreme pressure greases.

4. No correlation can be inferred between the results of the test and field service unless such correlation has been established.

Leakage Tendencies of Wheel Bearing Greases

ASTM Test for Leakage Tendencies of Automotive Wheel Bearing Greases (D 1263) is used to evaluate the leakage tendencies of wheel bearing greases during testing under prescribed laboratory conditions.

The test employs a modified automotive front hub assembly that is operated at 660 rpm for 6 h at 104°C (220°F). The smaller bearing is packed with 2 g of grease and the larger bearing with 3 g. Eighty-five grams of grease are put into the hub. After the test, the leakage in the hub-cap and leakage collector is determined. The bearings are washed and examined for varnish, gum, and lacquer-like material.

ASTM Test D 1263 provides a means to differentiate among grease products with distinctly different leakage characteristics. In addition, skilled operators can observe significant changes in other grease characteristics that may have occurred during the test. However, these observations are subject to differences in personal judgment and cannot be used for quantitative rating.

The test does not distinguish between wheel bearing greases having similar or borderline leakage.

Functional Life of Ball Bearing Greases

Two procedures for evaluating the functional life of ball bearing greases are contained in ASTM Test for Functional Life of Ball Bearing Greases (D 1741).

One procedure simulates the type of grease-packing commonly encountered in the field when the grease is pumped into the bearing housing with a grease gun. In the other procedure, one third of the free space in each bearing is filled with grease, while no grease is placed in the housing. This type of grease packing is typical of many applications including factory-packed bearings.

The tester used in D 1741 employs a pair of 30-mm (Type 306) deep-grooved radial ball bearings installed on a shaft. The bearings are run at 3500 rpm with a radial load of 11.2 kg (25 lb) and an axial load of 18.1 kg (40 lb) at temperatures up to 125°C (257°F). In both procedures, the bearings are run for 20 h, and then stopped and cooled for 4 h. This cycle is repeated until failure occurs or for a specified number of hours.

The test has significance because it may be used as a screening device to assist in the selection of greases which show promise as ball or roller bearing lubricants or both for use under the temperature limitations of the test. Close discrimination between lubricants is not possible with these procedures, and the method is limited to greases whose operating temperatures are not above 125°C (257°F). In addition, the method is not an equivalent to long service field tests.

Low-Temperature Torque

Greases harden and become more viscous as the temperature is lowered. In extreme cases the grease can become so rigid that excessive torque occurs within the bearing. Greases designed for extremely low temperatures must not stiffen nor offer excessive resistance to rotation.

ASTM Test for Low-Temperature Torque for Ball Bearing Greases (D 1478) measures the starting and running torques of lubricating greases packed in small bearings down to temperatures of −54°C (−65°F). In this procedure, fully packed bearings are installed on a spindle that can be rotated at 1 rpm. The assembly is inserted in a cold box. The outer race is connected by a string assembly to a spring scale where the restraining force is measured. When the motor is started, the initial peak restraining force is recorded. After running for 10 min, the restraining force is recorded again. These two values are multiplied by the length of the lever arm, and the products are reported as the starting and running torques in gram-centimeters (g-cm).

The test is significant because it is a means to compare the low-temperature torques of widely different greases. Since the method was developed using greases with extremely low-torque characteristics at −54°C (−65°F), it may not be applicable to other greases, speeds, or temperatures. Also, test conditions are substantially different from those found in the field so test results will not necessarily correlate with results from field tests.

Torque Stability, Wear, and Brine Sensitivity of Ball Joint Greases

ASTM Test for Torque Stability Wear and Brine Sensitivity Evaluation of Ball Joint Greases (D 3428) provides two procedures for evaluating the suitability of lubricating greases for use in automotive chassis ball joints.

The torque stability procedure provides a measure of the frictional and antiwear properties of lubricating greases when subjected to load and prolonged working under oscillating motions in ball joints.

The brine sensitivity test determines the ability of a lubricating grease to prevent noise from an oscillating ball joint when exposed to brine contamination.

In these tests, a grease lubricated automotive ball joint stud and bearing, confined in a fixed housing, is rocked through an arc under prescribed conditions of rocking frequency, rocking amplitude, load, and time. The noise, wear, and torque are observed or recorded or both.

The tests are used as a means to screen

lubricating greases to be used in automotive chassis ball joints. Since both the test components and procedures approximate those typical of current equipment and operation, test results are considered to provide some measure of grease suitability. Neither test is intended to measure the service life of an automotive chassis ball joint lubricating grease under actual service conditions.

Greases in Ball Bearings at Elevated Temperatures

ASTM Test for Performance Characteristics of Lubricating Greases in Ball Bearings at Elevated Temperatures (D 3336) is used to evaluate the performance characteristics of lubricating greases in ball bearings operating under light loads at high speeds and elevated temperatures.

With this test, the lubricating grease is evaluated in a 20-mm [Society of Automotive Engineers (SAE) No. 204] heat resistant steel ball bearing as it is rotated at 10 000 rpm under light load at a specified elevated temperature up to 371°C (700°F). The procedure provides for two cycle periods. One test cycle calls for 21½ h of operation on temperature and 2½ of shutdown without heat. The other test cycle calls for 20 h of operation on temperature and 4 h of shutdown without heat.

No correlation with actual field service can be assumed for this test.

Greases in Small Bearings

The computer and aircraft industries have used small bearings for many years. As the trend toward miniaturization mounted in other industries, a suitable test was needed to evaluate lubricating greases in small bearings. ASTM Evaluation of Greases in Small Bearings (D 3337) was developed to serve this purpose.

ASTM D 3337 may be used to determine grease life and torque in a small R-4 ball bearing. In this test the bearing is run at 12 000 rpm with a ½-lb radial and 5-lb axial load. If high-temperature bearings are used, the equipment is capable of testing up to 316°C (600°F).

The test has significance because it can be used as a means to differentiate between expected lives of greases used in small bearings at high speeds and temperatures. Grease torque also can be measured at both low and high speeds using this procedure.

The method will not differentiate between greases of closely related characteristics, and it should not be considered the equivalent of long-time, field service tests.

Grease Flow Properties at High Temperatures

The consistency of a grease is a critical parameter which defines the ability of a grease to perform in a fixed operating environment. The dropping point and oil separation tests define the temperature at which the thickener melts or if oil separates from the grease under high-temperature conditions. Neither of these evaluations indicate less dramatic, yet significant, changes in consistency as the grease approaches a fixed temperature.

ASTM Measurement of Flow Properties of Lubricating Greases at High Temperatures (D 3232) can be used to measure the flow properties of lubricating greases under high-temperature, low-shear conditions. Using this method, a grease sample is packed in an aluminum block. The packed block is placed on a hot plate capable of attaining temperatures in excess of 316°C (600°F) at a bearing rate of 5 ± 1°C (10 ± 2°F)/min. A special trident probe spindle, attached to a Brookfield viscometer, is lowered into the grease sample, and the hot plate is turned on. Simultaneously, the spindle is caused to rotate at a constant speed of 20 rpm. Torque measurements are read from the viscometer every minute. Readings are continued until the reading drops below 0.5 on the viscometer scale or until the maximum sample temperature of interest is attained. With this data and an appropriate conversion table, a plot of apparent viscosity versus temperature is prepared.

The results of this test provide a direct indication of the flow properties of a grease between room and elevated temperatures. Although the test does not give actual flow rates as in a pipeline, it is a means for obtaining some indication of this property.

Deleterious Particles in Lubricating Grease

ASTM Estimation of Deleterious Particles in Lubricating Grease (D 1404) defines a deleterious particle as one which will scratch a polished plastic surface. The test is applicable regardless of grease color or fillers and also may be used for testing other semisolid or heavy liquid substances.

With this method, the lubricating grease being tested is placed between two clean, highly polished acrylate plastic plates held rigidly and parallel to each other in metal holders. The assembly is pressed together squeezing the grease between the plastic plates to a thin layer. Any solid particles in the grease larger than the distance of separation of the plates and harder than the plastic will become imbedded in the opposing plastic surfaces. The apparatus is so constructed that one of the plates may be rotated about 30 deg with respect to the other while the whole assembly is under pressure. This will cause the imbedded particles to form characteristic arc-shaped scratches in one or both plates.

The relative number of such solid particles may be estimated by counting the total number of arc-shaped scratches on the two plates.

The test has significance because it is a rapid means for estimating the number of deleterious particles in a lubricating grease. However, a particle that is abrasive to plastic may not be abrasive to steel or other bearing materials. Therefore, the results of this test do not imply performance in field service.

Applicable ASTM/IP Standards

ASTM	*IP*	*Title*
D 217	50	Cone Penetration of Lubricating Grease
D 566	132	Dropping Point of Lubricating Grease
D 942	142	Oxidation Stability of Lubricating Greases by the Oxygen Bomb Method
D 1092		Apparent Viscosity of Lubricating Greases
D 1262		Lead in New and Used Greases
D 1263		Leakage Tendencies of Automotive Wheel Bearing Greases
D 1264		Water Washout Characteristics of Lubricating Greases
D 1402		Effect of Copper on Oxidation Stability of Lubricating Greases by the Oxygen Bomb Method (Intent to Withdraw)
D 1403	310	Cone Penetration of Lubrication Grease Using One-Quarter and One-Half Scale Cone Equipment
D 1404		Estimation of Deleterious Particles in Lubricating Grease
D 1478		Low-Temperature Torque of Ball Bearing Greases
D 1741		Functional Life of Ball Bearing Greases
D 1742		Oil Separation from Lubricating Grease During Storage
D 1743		Corrosive Preventive Properties of Lubricating Greases
D 1831		Roll Stability of Lubricating Grease

Applicable ASTM/IP Standards ***(continued)***

ASTM	*IP*	*Title*
D 2265		Dropping Point of Lubricating Grease Over Wide Temperature Range
D 2266		Wear Preventive Characteristics of Lubricating Grease (Four-Ball Method)
D 2509		Measurement of Extreme Pressure Properties of Lubricating Grease (Timken Method)
D 2595		Evaporation Loss of Lubricating Greases over Wide-Temperature Range
D 2596		Measurement of Extreme-Pressure Properties of Lubricating Greases (Four-Ball Method)
D 3232		Flow Properties of Lubricating Greases at High Temperatures
D 3336		Performance Characteristics of Lubricating Greases in Ball Bearings at Elevated Temperatures
D 3337		Evaluation of Greases in Small Bearings
D 3428		Torque Stability Wear and Brine Sensitivity Evaluation of Ball Joint Greases

10

Petroleum Waxes, Including Petrolatums

INTRODUCTION

IN THE EARLY YEARS OF PETROLEUM PROCESSING, the waxy materials that were separated were regarded as waste with no commercial value. Ultimately, however, wax products were recovered from the waste streams and, with the benefit of additional refining, were found to be useful replacements for natural waxes in many applications. A variety of grades of petroleum wax evolved that covered a broad range of physical properties. Now the petroleum wax product lines are in demand for a wide variety of uses. (It is of some interest to note that the end uses of wax are probably more diverse than the uses of any other petroleum product.)

Modern refining methods have made available select grades of wax of controllable and reproducible quality, and some have unique properties for specialized applications. Refined paraffin waxes, in grades varying by melting point and other qualities, represent the predominant segment of the total petroleum wax demand. The demand for tacky or ductile waxes and high-melting point waxes is met by the microcrystalline waxes selectively separated from the heavier fractions and residua of petroleum. In addition, complex blends of waxes have become increasingly important to the packaging industry, especially the hot melt blends containing petroleum wax base admixed with polymers, resins, or other additives.

OCCURRENCE AND REFINING OF WAXES

Wax is found in varying amounts in most crude oils. Paraffin base crudes (those rich in paraffin hydrocarbons) and mixed base crudes (containing both paraffins and naphthenes) usually have greater wax contents than aromatic base crudes. The type of wax recovered is dependent on the source and type of crude.

In the distillation of waxy crude oils, low-boiling cuts are removed first, followed by the intermediate and higher-boiling fractions which are suitable for making lubricating oils. To improve the low-temperature, pour-point property of lubricating oils, the lube cut is dewaxed. The by-product from dewaxing is refined to produce finished wax. Paraffin waxes are obtained from the intermediate boiling range lubricating oil fractions. Microcrystalline waxes are derived from the high-boiling lubricating oil fractions or residua.

Lube oil dewaxing in modern operations involves filtering a chilled solution of waxy lubricating oil in a specific solvent. The waxy material thus removed is designated as "slack wax." Slack wax contains a significant amount of oil that must be removed before the wax is suitable for most applications.

A solvent crystallization process generally is used for deoiling slack waxes. The raw wax is dissolved in solvent (for example, a methyl ethyl ketone-toluene mixture); the solution is chilled; and the crystallized wax is filtered out. Various grades of wax may be obtained by repeating the deoiling step at successively lower temperatures. The crystallized paraffin wax obtained from deoiling usually contains less than 0.5 percent oil. Some deoiled slack wax which is sold as crude scale wax contains about 2 percent oil.

The final stages of making refined wax are for purposes of improving the color, odor, and purity. Color and odor bodies and trace impurities of olefinic and aromatic hydrocarbons are removed by either hydrogenation or sulfuric acid treatment. In an alternative process the wax may be decolorized and deodorized by clay filtration. Following either procedure, the final product is termed a refined paraffin wax or a refined microcrystalline wax.

Petrolatums are oil-bearing products which are made from the heavy lubricating oil fraction in residua. Various processes are used to remove excessive oil, color, and odor bodies including combinations of such methods as solvent dilution and settling or centrifugation, fractional solvent crystallization, and percolation through clay or other adsorbent. The degree of refining varies—industrial grades of petrolatum require minimal processing, while pharmaceutical grades (the so-called petroleum jelly) require very selective treatment.

DEFINITIONS

A close classification of waxes is difficult because of the wide variety of waxes that have become available through the expansion of refining technology and the apparent overlapping among the types. However, refined petroleum waxes often are identified as paraffin wax, microcrystalline wax, or petrolatum.

Paraffin wax is petroleum wax consisting mainly of normal alkanes, varying amounts of condensed cycloalkanes, isoalkanes, and occasionally a very low percentage of aromatic material. Molecular weights are usually less than 450, and the viscosity at 98.9°C (210°F) normally will be less than 6 cSt. Either needle or plate type crystal structures are common. Paraffin waxes also exhibit pronounced latent heats of crystallization, and some have transition points (temperatures at which crystal structure modification occurs below the apparent solidification point of the wax).

Microcrystalline waxes contain substantial portions of hydrocarbons other than normal alkanes, and the components usually have higher molecular weights than paraffin wax components. Microcrystalline waxes also have smaller crystal structures and greater affinity for oil than paraffin waxes. Microcrystalline waxes usually melt between 65.6 and 104.4°C (150 and 220°F) and have viscosities between 10 and 20 cSt at 98.9°C (210°F).

A broad definition of petrolatum is: "A petroleum wax of unctuous structure containing substantial amounts of oil." The definition embraces the unrefined industrial grades as well as the refined pharmaceutical grades. The semisolid nature of petrolatum is due to the gel-like dispersion of a relatively high proportion of oil (usually above 10 percent) in the microcrystalline wax base. Petrolatums resemble microcrystalline waxes in composition, but contain lower proportions of normal paraffins. Melting points range from about 37.8 to 79.4°C (100 to 175°F) and viscosities range from 10 to 25 cSt at 98.9°C (210°F). Petrolatum colors range from the almost black crude form to the highly refined white pharmaceutical grade.

Petroleum wax is also an ingredient in certain so-called hot melts, and some test procedures may be applicable to hot melts. Hot melts have been defined as: "Those coatings, adhesive, or impregnants normally solid at ambient temperatures and liquid at elevated temperatures which are applied to substrates as hot liquids without the addition of volatile diluents." Wax-based hot melts are blends of wax, polymers, resins, or other additives. Hot melt products vary widely in total additive content depending on the end use. Many contain more than 25 percent additives. The additives serve to improve such properties as wax coating flexibility, barrier properties, gloss, scuff resistance, and viscosity.

Product quality and purity characteristics for medicinal uses of petrolatum are contained in the *U.S. Pharmacopeia.* Parrafin and microcrystalline waxes are simiiliarly defined in the *National Formulary.*

APPLICATIONS FOR WAX

Wax is one of the most versatile products of the oil industry. As a coating for a variety of paper, film, and foil substrates in the packaging industry, wax provides improved appearance, strength, moisture proofing, food oil resistance, sealing strength, and other desirable effects. The packaging industry consumes the major share of refined petroleum waxes either in the form of straight waxes or as hot-melt blends.

Wax has long served as a source of light and heat in such applications as candles, matches, and flares. In other uses, the high-gloss characteristic of some petroleum waxes makes them suitable ingredients for polishes. The highly refined waxes have excellent electrical properties for application in the insulation of small transformers, coils, capacitors, and certain cables. Waxes are used for the coating of fruit and cheese and the lining of cans and barrels. Blends of waxes are used by dentists for making dentures and engineers for mass-producing precision castings. Waxy petroleum streams also are used as feedstock for the production of numerous petrochemicals.

QUALITY CRITERIA

Criteria for judging the quality of a wax must be specific for the type of wax and its intended application.

Three general categories of properties are useful for assessing the quality of waxes: (*a*) physical properties (characterizations used for quality control); (*b*) chemical properties and composition (used for basic characterizations); (*c*) functional properties (empirical evaluations under simulated consumer conditions).

Criteria for Judging Physical Properties

Melting Point

The melting point is one of the most widely used tests to determine the quality and type of wax. Petroleum waxes usually do not melt at sharply defined temperatures because they are mixtures of hydrocarbons with different melting points. Paraffin waxes, as relatively simple mixtures, usually have narrower melting ranges than do microcrystalline waxes and petrolatums which are more complex in composition and melting behavior.

The melting point of a wax has direct and indirect significance in most applications. Paraffin waxes are usually marketed on the basis of melting point determined from the plateau in the cooling curve (ASTM D 87), which was once termed the English melting point (EMP). (Waxes are also sometimes marketed on the basis of so-called American melting point (AMP), which by definition is −16°C (3°F) above the EMP.) Melting point of a wax grade is often expressed as a range, that is, 130/135 MP.

Some waxes (especially narrow-cut, highly paraffinic types) undergo a phase change or transition point in the solid state which is usually about −3.9 to 4.4°C (25 to 40°F) below the melting point. Accompanying this change of crystalline form there may be significant changes in properties such as refractive index, density, flexibility, hardness, coefficients of expansion and friction, tensile strength, sealing strength, and gloss. For applications requiring consistency of properties over a specific temperature range, waxes are selected which do not exhibit a transition point in the critical range.

Viscosity

The fluidity of molten wax is important in applications involving coating or dipping processes, since it influences the amount of wax that is deposited on the substrate as well as the performance effectiveness of the coating obtained. Examples of such applications are in paper converting, hot-dip anti-corrosion coatings, and taper manufacturing.

Paraffin waxes do not differ widely in viscosity, typically falling in the range of 3 to 6 cSt at 98.9°C (210°F), although some grades have viscosities as high as 10 cSt. Microcrystalline waxes are considerably more viscous and vary over a wide range—about 10 to 20 cSt at 98.9°C (210°F).

The hot-melt blends of petroleum waxes with various additive modifiers cover a wide spectrum of viscosity. With additives of relatively low-molecular weight, only minor thickening effects occur. (Some of the resinous group of additives are prime examples.) On the other hand, additives that are high polymers, such as the polyethylenes or various copolymers, may have very high-thickening power. Thus, the components in a hot-melt formulation will be selected to give the proper viscosity consistent with the expected application conditions in the designated waxing equipment. Optimum viscosities usually exist for the various dipping operations, roll coaters, curtain coaters, or high-shear coaters. Viscosities of hot melts for coatings may range as high as 20 000 cP at 176.7°C (350°F), although 100 to 2000 cP is the more common range.

Hardness

Hardness is a measure of resistance to deformation, abrasion or damage; hence, it is an important criterion for many wax applications. Narrow-cut paraffin waxes generally have greater hardness than broad-cut waxes of the same melting point, and waxes with higher proportions of normal paraffins have greater hardness than highly branched hydrocarbons. Hardness is indirectly related to blocking tendency and gloss. Hard waxes usually have higher blocking points and better gloss than waxes of the same average molecular weight but wider molecular weight range.

Hardness is determined by measuring the depth of penetration of a needle or cone into a wax specimen. Techniques for measuring hardness (cone penetration for softer waxes and needle penetration for the harder waxes) have been standardized by American Society for Testing and Materials (ASTM) and Institute of Petroleum (IP) and will be described in later paragraphs.

Strength of Wax

Another quality test for wax is tensile strength. Some manufacturers and consumers find this to be a useful guide in monitoring the quality of the wax, although the tensile strength of a wax is extremely sensitive to the conditions of preparation of the wax specimen and the conditions of applying the tensile load.

Criteria for Judging Chemical Properties

Chemical properties and the composition of wax give a good indication of its degree of refinement. The properties of greatest significance are color, odor, oil content, amounts of solvent extractables, boiling point distribution, molecular weight, ultraviolet absorptivity, peroxide content, and amounts of carbonizable substances. In addition, there are specific chemical properties criteria for food grade wax.

Color

Paraffin waxes are generally of a translucent white color, whereas microcrystalline waxes and petrolatums range from white to almost black. A fully refined wax should be virtually colorless ("water-white") when examined in the molten state; typical scale wax usually exhibits a slight yellow or "straw" tint. Absence of color is of particular importance in waxes used for pharmaceutical purposes or the manufacture of food wrappings.

The acceptability of the color of microcrystalline waxes and petrolatums depends on the use for which they are intended. In some applications (for example, the manufacture of corrosion preventives), color may be of little importance, but in others it may be critical. For pharmaceutical purposes, the color of petrolatum is recognized in the grade specifications with colors ranging from white to yellow.

Colors in the solid state normally are expressed in descriptive terms such as white, off-white, yellow, amber, brown, etc. However, there is no standard test method for measuring this characteristic.

Odor

Freedom from odor and taste is particularly important in applications where the wax is likely to contact foodstuffs. When poor wax odor is detected, it may be due to inadequate refining, contamination during transport, storage or waxing operations, or deterioration in use. Odors due to inadequate refining are usually associated with a high-oil content or, less frequently, with residual solvent. Odor may be acquired during storage by absorption of odorous material from the container in which the wax slabs are packed. Wax also readily absorbs the odors of products such as cheese or soap if stored in proximity. Wax that has been overheated in use tends to oxidize and develop an odor similar to rancid coconut oil.

Subjective evaluations, such as odor, are difficult to standardize. There is, however, a technique which ensures reasonable concordance within a group whose members have agreed about odor level. Difficulties arise when there is a difference of opinion as to what constitutes an acceptable odor.

Oil Content

Because petroleum waxes are mixtures of hydrocarbons, which cover a range of molecular type and molecular weight, some components of a given wax may be of lower melting point and show more solubility or extractability than other fractions of the wax. These components are identified as the oil content of wax when they are determined as the soluble fraction in the methyl ethyl ketone crystallization of the oil content method [ASTM Oil Content of Petroleum Waxes (D 721/IP 158)]. It should be recognized that there is no clear-cut division between oil and wax with regard to particular chemical species.

It is necessary to distinguish between paraffin wax, microcrystalline wax, and petrolatums when considering the significance of oil content because of the different degree of affinity that each of these wax types has for oil.

Fully refined paraffin wax usually has an oil content of less than 0.5 percent. Waxes containing somewhat more than this amount of oil (that is, 0.5 to about 2.0 percent) are referred to as "scale wax." An intermediate grade known as "semirefined wax" is sometimes recognized for waxes having an oil content of about 1 percent. Excess oil tends to exude from paraffin wax giving it a dull appearance and a greasy feel. A high-oil content tends to have an adverse effect on sealing strength, tensile strength, hardness, odor, taste, color, and color stability. High-oil content also tends to plasticize wax and increase its flexibility.

Microcrystalline waxes have a greater affinity for oil than paraffin waxes because of their smaller crystal structure. The permissible amount depends on the type of wax and its intended use. The oil content of microcrystalline wax is, in general, much greater than that of paraffin wax and could be as high as 10 percent. Waxes containing more than 10 percent oil would usually be classed as petrolatums, but the demarcation is by no means precise.

Solvent Extractables

For waxes containing no more than 15 percent oil, ASTM Method D 721/IP 158 is used to define oil content. For waxy streams that contain more than 15 percent oil and various "soft" or soluble fractions (for example, petrolatum and slack waxes), a new procedure has been developed [ASTM Test for Solvent Extractables in Petroleum Waxes (D 3235)].

(It should be remembered that the solvent separations are empirical, and the distinction between "oil content" and "extractables" is associated with the different solvent compositions. The terminology should not be used interchangeably. Thus, it is erroneous to indicate an oil content as being determined by ASTM D 3235.)

Boiling Point Distribution

Boiling point distribution is useful because it provides an estimate of hydrocarbon molecular weight distribution that influences many of the physical and functional properties of petroleum wax. (To a lesser extent, distillation characteristics also are influenced by the distribution of various molecular types; that is, normal paraffins, branched, or cyclic structures. In the case of the paraffin waxes which are predominantly straight chain, the distillation curve reflects the molecular size distribution.)

In the most common distillation test, ASTM Test for Distillation of Petroleum Products at Reduced Pressure (D 1160), cuts are obtained under reduced pressure, such as at 10 mm for paraffin waxes or at 1 mm for higher molecular weight waxes. The cuts are taken at intervals across the full distillable range, and the complete results may be reported. In some cases, and for brevity, the distillation results are reported as the temperature difference observed between the 5 percent off and 95 percent off cut points. This would represent a "width of cut" statement, expressed in degrees, at the reduced pressure. Waxes having very narrow width of cut will tend to be more crystalline, have higher melting points, higher hardness and tensile strength properties, and less flexibility.

Molecular Weight

The molecular weight of various waxes may differ according to: (1) the source of the wax (whether it originated in the lighter or heavier grade lubricating oils), and (2) the processing of the wax (the closeness of the distillation cut or the fractionation by crystallization). Thus, the average molecular weight of a wax may represent an average of a narrow or a wide band of distribution.

As a generalization, for any series of similar waxes, an increase in molecular weight increases viscosity and melting point. However, many of the other physical and functional properties are more related to the hydrocarbon types and distribution than to the average molecular weight.

The procedure for determination of number average molecular weight is described in later paragraphs.

Criteria for Food Grade Wax

Federal regulation 21 CFR 172.886 contains a specification that describes food grade wax as a "mixture of solid hydrocarbon, paraffinic in nature, derived from petroleum and refined to meet specifications."

Besides normal guidelines, this U.S. specification and its counterpart in the United Kingdom impose stringent controls on the polynuclear aromatic hydrocarbon content of food grade waxes for direct use in foods or indirect contact with foods in packaging materials. The controls are to ensure the essential absence of polynuclear hydrocarbons, especially those that may be carcinogenic. The rigorous procedure for estimating polynuclear aromatic content of petroleum waxes is described in the section of this chapter on tests for chemical properties.

Ultraviolet Absorptivity

For some process control purposes or for research purposes, it may be desirable to monitor the total aromatic content of a petroleum wax. ASTM Test for Ultraviolet Absorbance and Absorptivity of Petroleum Products (D 2008) provides this capability.

Although this procedure shows good operator precision, the interpretation of results requires some caution. Since the test does not include any selective fractionation of the sample, it does not distinguish any particular aromatic. It is also subject to the errors arising from interferences or differences in strong or weak absorptivities shown by different aromatics. Therefore, the test is good for characterization, but cannot be used for quantitative determination of aromatic content or any other absorptive component.

Carbonizable Substance

The so-called "Readily Carbonizable Substance" test is no longer in wide use, although it is still a requirement for paraffin wax in the monograph in the *U.S. National Formulary,* Volume 16 (U.S.P./N.F.: XXI/XVI). The test has some utility as a check on the degree of removal of trace reactive materials in the finishing of the wax products, but it is subject to errors and misleading interpretation. In addition, it does not correlate with specific impurities of interest, nor is it a good quantitative measure.

Peroxide Content

The presence of peroxides or oxidized products such as aldehydes or fatty acids are found in wax as a result of oxidation and deterioration of wax either in use or storage. Suitable antioxidants, such as 2,6 di-tertiary butyl-p-cresol and butylated hydroxyanisole, may be used to retard oxidation. ASTM Test for Peroxide Number of Petroleum Wax (D 1832) is used to measure the peroxide content of wax.

Criteria for Judging Functional Properties

Testing methods for evaluating functional properties are often empirical in nature and based on the simulated conditions of use. The results of such tests are intended to correlate with practice and used to determine the suitability of a wax for a particular application.

The properties of waxed paper or board depend not only on the individual characteristics of the wax and paper components, but also on the manner in which they interact when they are combined in the waxing process. For this reason, it is necessary to conduct certain functional tests on the finished paper.

According to the method of application, the wax may be coated on one or both surfaces of the paper, or it may be impregnated into the pores and fibers of the paper.

Wax Content of Substrates

Many of the functional properties of coated board or waxed paper are dependent on the amounts of wax present either on the surface or internally.

Surface wax on each or both sides of a weighed specimen can be determined by scraping each side and weighing the specimen after each operation. Internal wax is the difference between total wax content and total surface wax. Total wax content can be determined by solvent extraction or by finding the difference between average weight per unit area of waxed and unwaxed specimens.

Blocking Properties

Blocking tendency is an important quality criterion for waxes, especially those used in various paper and paperboard coating applications. Blocking occurs at moderately warm temperatures when waxed surfaces stick together or block. If the surfaces of waxed paper block, the surface appearance, gloss, and barrier properties are destroyed when the papers are separated. The blocking and picking points indicate the temperature range at which waxed film surfaces become damaged if contact is made.

In general, a low resistance to blocking is associated with the presence of low-molecular weight and nonnormal fractions in the wax, such as oil or soft waxes. Likewise, broad-cut waxes generally show somewhat lower blocking points than shown by narrow-cut waxes of the same average molecular weight or viscosity. In addition, the blocking point of a given wax may also be dependent on the conditions under which the particular wax coating is prepared. For example, subtle changes in the crystalline nature of the coating can occur with variations in paper substrate quality, coating application temperature, chilling conditions to set the coating, and particulars of conditioning or aging of the coating. Standard testing procedures are designed to control these factors.

Testing for blocking tendency is done usually with a moderate load (0.3 psi) applied to two adjacent waxed paper surfaces which are exposed to the range of temperature across a gradient heating plate. The test serves to simulate normal packaging quality needs.

Gloss

Waxed coatings not only provide protection for packaged goods, but they also provide an attractive appearance because of their high-gloss characteristics. The amount of gloss obtained is determined primarily by the nature of the wax and coating and the smoothness of the substrate. Storage of coated paper at elevated temperature tends to be injurious to gloss. Measurement of gloss properties and gloss retention are described in later paragraphs.

Slip Properties

Friction tests provide an indication of the resistance to sliding exhibited by two waxed surfaces in contact with one another. The intended application determines the degree of slip that is desired. Coatings for packages that require stacking should have a high coefficient of friction to prevent slippage in the stacks. Folding box coatings should have a low coefficient of friction to allow the boxes to slide easily from a stack of blanks being fed to the forming and filling equipment.

Abrasion Resistance

The attractive appearance of a clear, glossy wax coating is an important functional feature in many packaging applications such as the multicolored food wraps, folding cartons, and decorated, corrugated cartons. However, the surface appearance may be vulnerable to damage if its handling in manufacture or use subjects it to any severe abrading action. A standard procedure has been set up to evaluate the abrasion resistance of a wax coating under impact of a stream of sand and using gloss change as the criterion for evaluating abrasion.

Adhesion

The thermoplastic properties of wax are used to good effect in the heat sealing of waxed paper packages. The strength of the seal is a function not only of the physical properties but also of the chemical properties and composition of the wax.

In some applications, such as for bread wraps, candy wraps, cereal, and cookie inner bags, it is desirable to have only a moderately strong seal that can be parted easily when the package is opened without any tearing action of the paper itself. For these uses, predominantly paraffinic wax blends are used, and seal testing shows a cohesive failure of the wax. In other applications, such as for various heat-sealed folding cartons, frozen food cartons, and flexible packaging bags, extremely strong and durable seals may be desired. In these cases, the wax formulation contains a major proportion of tacky, adhesive, and high-tensile components such as the resins and polymers compounded in hot melts. During seal testing the wax cohesive and adhesive strength may be so great that the failure occurs in the paper substrate rather than in the wax.

Moisture Barrier Properties

The ability of wax to prevent the transfer of moisture vapor is a primary concern to the food packaging industry. Moisture must be kept out to maintain the freshness of dry foods, and moisture must be kept in to maintain the quality of frozen foods and baked goods. Both require different procedures for measuring barrier properties. For example, for dry food packaging, transmission rate is measured at elevated temperatures and high-relative humidity, and, for frozen food, the transmission rate is measured at low temperatures and low-relative humidity.

To have good barrier properties, a wax must be applied in a smooth, continuous film and be somewhat flexible to prevent cracking and peeling. Testing in the "flat" condition provides an assessment of the fundamental barrier capability of the wax coating; for example, the resistance of the microstructure to passage of water vapor. Testing in the "creased" condition gives an evaluation of the strength of the thin wax film in maintaining its integrity even when folded to a sharp crease through 180 deg and reflattened for test. A strong, flexible, adherent film will survive the creasing with minimal damage, while inferior coatings will lose integrity, crack, or flake off at the crease.

TEST METHODS

Quality criteria for wax are a product of the type wax and its intended application. Three categories of tests—physical, chemical, and functional—are used to assess the quality of waxes. The following paragraphs describe the most common tests applied to waxes and petrolatums.

Physical Properties of Petroleum Wax

ASTM Test for Melting Point of Petroleum Wax (Cooling Curve) (D 87/IP 55)—In this procedure, a sample of molten paraffin wax in a test tube, with a thermometer, is placed in an air bath which is inserted into a room temperature water bath. As the wax cools, wax temperature versus time is recorded and plotted. Since the latent heat of crystallization released during solidification of the wax is sufficient to temporarily arrest the rate of cooling, a plateau occurs in the time/temperature curve. The temperature at which the plateau occurs is the melting point. This procedure is not suitable for microcrystalline wax, petrolatums, or waxes containing large amounts of nonnormal hydrocarbons because the plateau rarely occurs in cooling curves of such waxes.

ASTM Test for Drop Melting Point of Petroleum Wax, Including Petrolatum (D 127/IP 133)—This method can be used for most petroleum waxes and wax-based hot-melt blends. A chilled thermometer bulb is coated with the molten wax (which is allowed to solidify), placed in a test tube, and heated at a specified rate in a water bath. The melting point is the temperature at which the first drop of liquid falls from the thermometer.

ASTM Test for Congealing Point of Petroleum Wax, Including Petrolatum (D 938/IP 76)—This procedure can be used for almost all types of petroleum waxes and hot-melt blends. A thermometer bulb is dipped in the melted wax and placed in a heated flask, serving as a jacket. The thermometer is held horizontally and slowly rotated on its axis. As long as the wax remains liquid, it will hang from the bulb as a pendant drop. The temperature at which the drop rotates with the thermometer is the congealing point. The congealing point of a microcrystalline wax or petrolatum is invariably lower than its drop melting point.

ASTM Test for Kinematic Viscosity of Transparent and Opaque Liquids (and the Calculation of Dynamic Viscosity) (D 445/IP 71)—Kinematic viscosity is measured by timing the flow of a fixed volume of material through a calibrated capillary at a selected temperature. The unit of kinematic viscosity is the stoke, and kinematic viscosities of waxes are usually reported in centistokes. Centistokes can be converted to Saybolt Universal seconds by using ASTM Conversion of Kinematic Viscosity to Saybolt Universal Viscosity or to Saybolt Furol Viscosity (D 2161).

ASTM Test for Apparent Viscosity of Petroleum Waxes with Additives (Hot Melts) (D 2669)—This method is suitable for blends of wax and additives having apparent viscosities up to 20 000 cP at 176.7°C (350°F). Apparent viscosity is the measurement of drag produced on a rotating spindle immersed in the test liquid. A suitable viscometer is equipped to use interchangeable spindles and adjustable rates of rotation. The wax blend is brought to test temperature in a simple arrangement using an 800-ml beaker in a heating mantle. Viscosities over a range of temperatures may be recorded and plotted on semilog paper to determine the apparent viscosity at any temperature in the particular region of interest.

ASTM Test for Apparent Viscosity of Hot Melt Adhesives and Coating Materials (D 3236)—This procedure is similar to ASTM D 2669 since it measures the ratio of shear stress to shear rate through the use of a rotating spindle. However, ASTM D 3236 is applicable to hot melts of a broader range of viscosity. Through the use of a special thermally controlled sample chamber, hot melts may be tested at temperatures of 176.7°C (350°F) or more and at viscosities of 200 000 cP or more.

ASTM Test for Needle Penetration of Petroleum Waxes (D 1321)—The hardness or consistency of wax is measured with a penetrometer applying a load of 100 g for 5 s to a standard needle having a truncated cone tip. To prepare a wax for testing, the sample is heated to 16.7°C (30°F) above its congealing point, poured into a small brass cylinder, cooled, and placed in a water bath at the test temperature for 1 h. The sample is then positioned under the penetrometer needle which when released penetrates into the sample. The depth of penetration in tenths of a millimeter is reported as the test value. This method is not applicable to oily materials or petrolatums which have penetrations greater than 250.

ASTM Test for Cone Penetration of Petrolatums (D 937/IP 179)—The Cone Penetration test is used for soft waxes and petrolatums. It is similar to D 1321 except that a much larger sample mold is used and a cone weighing 102.5 g replaces the needle. The method requires that a 150 g load be applied for 5 s at the desired temperature.

ASTM Test for Tensile Strength of Paraffin Wax (D 1320)—This test is an empirical evaluation of the tensile strength of waxes which do not elongate more than 1/8 in. under the test conditions. Six dumbbell-shaped specimens, each with a cross sectional area of 1/4 in.^{2} at the neck of the dumbbell, are cast. The specimens are broken on a testing machine under a load which increases at the rate of 20 lb/s along the longitudinal axis of the sample. Values are reported as pounds per square inch.

Chemical Properties of Petroleum Wax

ASTM Test for Saybolt Color of Petroleum Products (Saybolt Chromometer Method) (D 156)—Saybolt color is determined on nearly colorless waxes by putting the melted sample in a heated vertical tube mounted alongside a second tube containing standard color discs. An optical viewer allows simultaneous viewing of both tubes. Light is reflected by a mirror up into the tubes and to the viewer. The level of the sample in the column is decreased until its color is lighter than that of the standard. The color number above this level is reported.

ASTM Color of Petroleum Products (ASTM Color Scale) (D 1500/IP 196)—This procedure is used for waxes and petrolatums which are too dark for the Saybolt colorimeter. Using a standard light source, a liquid sample is placed in the test container and compared with colored glass discs ranging in value from 0.5 to 8.0. If an exact match is not found and the sample color falls between two standard colors, the higher of the two colors is reported.

ASTM Test for Odor of Petroleum Wax (D 1833/IP 185)—The odor of petroleum wax is determined by a preselected panel. Ten grams of wax are shaved and placed in an odor-free glass bottle and capped. After 15 min, the sample is evaluated in an odor-free room by removing the cap and sniffing lightly. A rating of 0 (no odor) to 4 (very strong odor) is given by each member of the panel. The reported value is the average of the individual ratings.

ASTM Test for Oil Content of Petroleum Waxes (D 721/IP 158)—The fact that oil is much more soluble than wax in methyl ethyl ketone at low temperatures is utilized in this procedure. A weighed sample of wax is dissolved in warm methyl ethyl ketone in a test tube and chilled to −31°C (−25°F) to precipitate the wax. The solvent-oil solution is separated from the wax by pressure filtration through a sintered glass filter stick. The solvent is evaporated, and the oil residue is weighed. The method is applicable to waxes containing not more than 15 percent oil.

ASTM Test for Solvent Extractables in Petroleum Wax (D 3235)—This method is very similar to ASTM D 721/IP 158 except that the solvent used is a 1:1 mixture of methyl ethyl ketone and toluene. The method may be used on waxes having high levels of extractables, for example, 15 to 50 percent extractables.

ASTM Test for Distillation of Petroleum Products at Reduced Pressure (D 1160)—In this standard test the sample is distilled at an accurately controlled, reduced pressure (usually 10 mm or 1 mm) in a distillation flask and column which provides approximately one theoretical plate fractionation. The apparatus includes a condenser and a receiver for collection and measurement of the volume of cuts obtained. The data are reported as the series of vapor temperatures observed (at the reduced pressure) for the specified intervals of volumetric percentage distilled.

ASTM Test for Boiling Range Distribution of Petroleum Fractions by Gas Chromatography (D 2887)—A sample of the test wax is dissolved in xylene and introduced into a gas chromatographic column which is programmed to separate the hydrocarbons in boiling point order by raising the temperature of the column at a reproducible, calibrated rate. When wax samples are used, the thermal conductivity detector is used to measure the amount of eluted fraction.

Data obtained in this procedure are reported in terms of percent recovered at certain fixed temperature intervals. A full description of the carbon number distribution cannot be obtained with this method because the relatively short column used does not provide sufficient resolution to distinguish individual peaks. In addition, the method is not applicable to higher chain length paraffin waxes or to microcrystalline waxes.

ASTM Test for Molecular Weight of Hydrocarbons by Thermoelectric Measurement of Vapor Pressures (D 2503)—In this method, a small sample of wax is dissolved in a suitable solvent, and a droplet of the wax solution is placed on a thermistor in a closed chamber in close proximity to a suspended drop of the pure solvent on a second thermistor. The difference in vapor pressure between the two positions results in solvent transport and condensation onto the wax solution with a resultant change in temperature. Through suitable calibration, the observed effect can be expressed in terms of molecular weight of the wax specimen as a number average ($\bar{M}_n$).

Regulatory Test for Food Grade Petroleum Wax: 21 Code of Federal Regulations 172.886 and U.K. Mineral Hydrocarbons in Food Regulations—The polynuclear aromatic content of waxes can be estimated by the ultraviolet absorbance of an extract of the sample when scanned in the 280 to 400 nm range. The sample is first dissolved in *iso*octane and extracted using a dimethylsulfoxide-phosphoric acid solution to concentrate the aromatic material and remove it from the saturated hydrocarbons. The extract portion is tested for ultraviolet absorbance against specified maximum limits. These limits were established as being indicative of the maximum absorption peak areas associated with polynuclear aromatic hydrocarbons. If the wax sample under test falls within these limits, the sample is considered to pass, and no further continuation of the analysis is necessary.

If, however, the results in this first phase of testing exceed any of the limits, there is a possibility that the high result may be an interference due to single ring aromatic derivatives (these are highly absorptive, especially at lower wavelengths and would normally occur along with the polynuclear aromatics). The benzenoid structures are considered noncarcinogenic, and their interference in the test must be eliminated. This is done by continuing the sample extract into a second phase of analysis using a chromatographic separation procedure. The resulting concentrate of polynuclear aromatics is tested against the same specified limits for ultraviolet absorbance.

ASTM Test for Ultraviolet Absorbance and Absorptivity of Petroleum Products (D 2008)—ASTM D 2008 is a standard method for testing a variety of petroleum products of liquid or solid form. The procedure tests the product as a whole, without including any separation or fractionation steps to concentrate the absorptive fractions. When wax or petrolatum are tested in this procedure, the specimen is dissolved in *iso*octane, and the ultraviolet absorbance is measured at a specified wavelength such as 290 nm. The absorptivity is then calculated. This procedure, as such, is not a part of the federal specification.

ASTM Test for Carbonizable Substances in Paraffin Wax (D 612)—Five millilitres of concentrated sulfuric acid are placed in a graduated test tube, and 5 ml of the melted wax are added. The sample is heated for 10 min at 70°C (158°F). During the last 5 min, the tube is shaken periodically. The color of the acid layer is compared with the color of a standard reference solution. The wax sample passes if the color is not darker than the standard.

ASTM Test for Peroxide Number of Petroleum Wax (D 1832)—In this test, a sample is dissolved in carbon tetrachloride, acidified with acetic acid, and a solution of potassium iodide is added. Any peroxides present will react with the potassium iodide to liberate iodine which is then titrated with sodium thiosulphate.

Functional Properties of Petroleum Wax

In the manufacture of wax-treated paper packaging, a variety of substrates are treated, and a variety of waxing methods and equipment are used. In some cases, commercial specimens of waxed packaging materials may be submitted to functional tests. Such testing serves to evaluate the total system, that is, the wax, paper, and coating condition.

When it is desired, to test the functional property of the wax itself, or to compare it with other waxes apart from the variables of substrate and waxing process, a standard waxed specimen must be prepared under predetermined and controlled conditions. Several of the functional test methods include instructions for specimen waxing which usually employ a standard paper substrate and standard coating conditions.

ASTM Test for Surface Wax on Waxed Paper or Paperboard (D 2423)—This method determines the amount of wax present as a surface film on a substrate but not the absorbed wax. A waxed paper sample is cut to size and weighed; wax is carefully scraped from one surface with a razor blade, then the sample is reweighed to determine the amount of wax removed. The process is repeated on the reverse side if total surface wax is desired.

ASTM Test for Total Wax Loading of Corrugated Paperboard (D 3344)—In this method, taking together all wax that may be present as impregnating or saturating or coating wax, the total wax loading of corrugated board is measured. The determination is made by extracting the waxed board specimen with warm solvent (1,1,1 trichlorethane) and evaporating to dryness to obtain extracted wax. The resulting residue will include any soluble additive materials associated with the wax, such as the polymeric additives normally used in the wax-based hot-melt coatings.

ASTM Test for Blocking and Picking Points of Petroleum Wax (D 1465)—Two strips of wax coated paper, 1-in. wide, are placed face to face between two uncoated strips of paper on a calibrated temperature-gradient heating plate. The specimens are covered with foam rubber strips and steel bars to provide a moderate pressure loading and then subjected to the gradient heat of the blocking plate for 17 h. The specimens are removed, cooled, and peeled apart. The picking point is the lowest temperature at which the surface film shows disruption. The blocking point is the lowest temperature causing disruption of surface film over an area of 50 percent of the strip width.

ASTM Test for 20-Deg Specular Gloss of Waxed Paper (D 1834)—Specular gloss is the capacity of a surface to simulate a mirror in its ability to reflect an incident light beam. The glossimeter used to measure gloss consists of a lamp and lens set to focus an incident light beam 20 deg from perpendicular to the specimen. A receptor lens and photocell are centered on the angle of reflectance, also at 20 deg from the perpendicular. A black, polished glass surface is used for instrument standardization at 100 gloss units. The wax coated paper specimen is held by a vacuum plate over the sample opening. The light beam is reflected from the sample surface into the photocell, the gloss being reported on a percentage basis.

ASTM Test for Gloss Retention of Waxed Paper and Paperboard after Storage at 40°C (104°F) (D 2895)—The gloss is measured by ASTM Method D 1834 before and after aging the sample for 1 and 7 days in an oven at 40°C (104°F). Gloss retention is the percent of original gloss retained by the specimen after aging under specified conditions. The aging conditions are intended to correlate with the conditions likely to occur in the handling and storage of waxed paper and paperboard.

ASTM Test for Coefficient of Kinetic Friction for Wax Coatings (D 2534)—A wax coated paper is fastened to a horizontal plate attached to the lower, movable cross arm of an electronic load-cell-type tension tester. A second paper is taped to a 180-g sled which is placed on the first sample. The sled is attached to the load cell through a pulley. The kinetic coefficient of friction is calculated from the average force required to move the sled at 35 in./min (90 cm/min) divided by the sled weight.

ASTM Test for Abrasion Resistance of Wax Coatings (D 3234)—The abrasion resistance of a smooth wax coating on paper or paperboard is measured by determining how much change in gloss of the surface occurs under the abrading action of sand particles. A measured quantity of sand is allowed to impinge under specified conditions on the waxed surface, and the change in gloss is noted using ASTM Method D 1834.

Water Vapor Permeability of Paper and Paperboard (TAPPI T 448)—The method is used for determining the water vapor permeability of wax treated substrates at normal atmospheric conditions: namely, 22.8°C (73°F) and 50 percent relative humidity. A dessicant is placed in a shallow test dish, and the specimens of wax coated paper are sealed to the shoulder of the dish above the dessicant to completely seal the dessicant in the dish. The assembly is weighed at the start and placed in the con-

TABLE 1. Inspections on examples of petroleum waxes.

Wax grade	127/129 AMP paraffin	156/158 AMP paraffin	150 microcrystalline	175 microcrystalline	USP petrolatum
Appearance	Translucent white solid	Translucent white solid	opaque white solid	yellow solid	white semi-solid
Gravity, °API, calculated	43.4	38.7	36.8	34.5	32.4
Melting point, ASTM D 87/IP 55, °F (°C)	125.2 (52)	154.0 (68)	...	...	...
Melting point, ASTM D 127/IP 133, °F (°C)	128 (53)	156 (69)	149 (65)	176 (80)	125 (53)
Congealing point, ASTM D 938/IP 76, °F (°C)	125 (52)	152 (67)	145 (63)	166 (74)	117 (47)
Penetration, ASTM D 1321 77°F, 100 g, 5 s	14	15	29	17	...
Penetration, ASTM D 937/IP 179 77°F, 150 g, 5 s	...	...	...	...	155
Color, Saybolt, ASTM D 156	30+	30+	14	...	8
Color, ASTM D 1500/IP 196	...	...	...	L1.0	L1.0
Viscosity, ASTM D 445/IP 71; cSt at 210°F	3.18	7.36	10.56	16.64	10.53
Refractive index, ASTM D 1747	1.432	1.444	1.450	1.456	1.45
Odor, ASTM D 1833/IP 185	1.0	1.0	1.5	2.0	1.5
Carbonizable substance, ASTM D 612	pass	...	...	...	pass
Oil content, ASTM D 721/IP 158 weight, %	0.2	0.1	2.4	0.1	50[a]
Specular gloss, ASTM D 1834	35	39	44	38	...
Blocking point, ASTM D 1465, °F (°C)	95 (35)	117 (47)	120 (49)	127 (53)	...
Water vapor permeability, TAPPI 464 Grams water/24 h/m^2					
Flat 26 g/ft^2 (m^2) deposition	6.5 (70)	8.0 (86)	...	4.2 (45)	...
Creased 32 g/ft^2 (m^2) deposition	9.7 (104)	36 (387)	...	12 (129)	...
Coefficient of friction, ASTM D 2534	0.26	1.0	0.92	0.64	...
Tensile strength, ASTM D 1320, psi (g/cm^2)	290 (2×10^4)	360 (2.5×10^4)	385 (2.7×10^4)	385 (2.7×10^4)	...

[a]Methyl isobutylketone used in place of methyl ethyl ketone for determination of oil content.

trolled testing room or cabinet at the conditions required. The dish assembly is reweighed at periodic intervals such as 24, 48, or 72 h intervals until successive weighings show that a constant rate of gain has been attained. The water vapor transmission rate (WVTR) is reported as the grams of water transmitted per square metre of test specimen per 24 h. The report includes notation as to whether the specimen was flat or creased before testing.

Water Vapor Permeability of Sheet Materials at High Temperature and Humidity (TAPPI T 464)—This method is similar to the T 464 method previously described, but evaluates the barrier properties of a waxed substrate under more severe conditions; namely, in an environment of 37.8°C (100°F) and 90 percent relative humidity.

INSPECTIONS OF TYPICAL PETROLEUM WAXES

Table 1 lists a number of inspection data on examples of petroleum waxes. These data are simply illustrative and presented for general information only. They are not to be regarded as specifications or limiting guidelines. Data for other wax samples of the same nominal grades may differ significantly from those quoted in Table 1.

Applicable ASTM/IP Standards

ASTM	*IP*	*Title*
D 87	55	Melting Point of Petroleum Wax (Cooling Curve)
D 127	133	Drop Melting Point of Petroleum Wax, Including Petrolatum
D 156		Saybolt Color of Petroleum Products (Saybolt Chromometer Method)
D 445	71	Kinematic Viscosity of Transparent and Opaque Liquids (and the Calculation of Dynamic Viscosity)
D 612		Carbonizable Substances in Paraffin Wax
D 721	158	Oil Content of Petroleum Waxes
D 937	179	Cone Penetration of Petrolatum
D 938	76	Congealing Point of Petroleum Waxes, Including Petrolatum
D 1160		Distillation of Petroleum Products at Reduced Pressure
D 1320		Tensile Strength of Paraffin Wax
D 1321		Needle Penetration of Petroleum Waxes
D 1465		Blocking and Picking of Petroleum Wax
D 1500	196	ASTM Color of Petroleum Products (ASTM Color Scale)
D 1832		Peroxide Number of Petroleum Wax
D 1833	185	Odor of Petroleum Wax
D 1834		20-Degree Specular Gloss of Waxed Paper
D 2008		Ultraviolet Absorbance and Absorptivity of Petroleum Products
D 2423		Method for Surface Wax on Waxed Paper or Paperboard
D 2534		Coefficient of Kinetic Friction for Wax Coatings
D 2669		Apparent Viscosity of Petroleum Waxes Compounded with Additives (Hot Melts)
D 2887		Boiling Range Distribution of Petroleum Fractions by Gas Chromatography
D 2895		Gloss Retention of Waxed Paper and Paperboard After Storage at 40°C (104°F)

Applicable ASTM/IP Standards ***(continued)***

ASTM	*IP*	*Title*
D 3234		Abrasion Resistance of Wax Coatings
D 3235		Solvent Extractables in Petroleum Waxes
D 3236		Apparent Viscosity of Hot Melt Adhesives and Coating Materials
D 3344		Total Wax Loading of Corrugated Paperboard

Bibliography

Bennet, H., *Industrial Waxes, Vol. I & II,* Chemical Publishing Company, Inc., New York, 1963.

Petroleum Waxes: Characterization, Performance, and Additives, Technical Association of the Pulp and Paper Industry, Special Technical Association Publication No. 2, 1963.

Technical Evaluation of Petroleum Waxes, Technical Association of Pulp and Paper Industry, Special Technical Association Publication No. 6, 1969.

The United States Pharmacopeia XXI/The National Formulary XVI, United States Pharmacopeial Convention, Inc., Rockville, MD.

Warth, A. H., *Chemistry and Technology of Waxes,* Reinhold Publishing Corp., New York, 1956.

White Mineral Oils

11

INTRODUCTION

"Russian Oil," originally made from the straw colored crude oil produced in the Surakhany district of Russia and first introduced abroad in 1913 for its therapeutic value in the treatment of constipation, was the grandfather of the white oils of today. Within the subsequent time span, many new processes and process improvements have been made to produce these unique and valuable petroleum products. At the same time, there has been an increasing demand by both government and industry for more stringent purity requirements and special properties for use in new applications.

Usage has expanded and, with it, the scope of products that may be classified as white mineral oils. In the context employed in this chapter, white mineral oils are considered to be those petroleum fractions, from kerosine to viscous lubricating oil, that have been drastically treated to remove all reactable hydrocarbons to give pure, colorless, odorless, and nonreactable petroleum products. It is the combination of these four properties, unique to these refined natural petroleum fractions, that has expanded the applications of white oils from a simple medicinal to their widespread use in such diverse fields as cosmetics, pharmaceuticals, plastics, agricultural and animal sprays, food processing and protection, animal feed, refrigeration and electrical equipment, chemical reagent and reaction media, and precision instrument lubricants.

MANUFACTURE

White oils are made from selected petroleum fractions distilled to provide finished products with the viscosity desired by the white oil user. For example, a manufacturer may have in his base white oil line, oils with viscosities of 4, 8, 12, 19, and 70 cSt (all at 40°C) (40, 50, 70, 100, and 370 SUS at 100°F). From these base oils the refiner is able to blend any supply white oils covering the viscosity range 4 to 70 cSt at 40°C (40 to 370 SUS at 100°F). In addition to viscosity, properties considered during feedstock selection include pour point, cloud point, distillation range, smoke point, specific gravity, etc.

Typically, a white oil feedstock is distilled to specification from the required crude oil cuts, pretreated by solvent extraction or hydrotreating, and dewaxed if necessary. The order of the feedstock preparation is dependent on the refiner's operations. The type pretreatment used is determined by the nature of the white oil manufacturer's process and the type base white oils that best fit his market.

With the advent of hydrotreating, white oils can be produced without acid treatment. However, in traditional manufacture, white oils are made by exhaustively treating the selected feedstock with 20 percent fuming sulfuric acid to remove the aromatic, unsaturated hydrocarbons and other impurities present in the natural petroleum oil. Gaseous sulfur trioxide treatment also may be used to remove reactable hydrocarbons. Petroleum sulfonic acids, produced as by-products during acid treatment, are removed by extraction and neutralization. The oil is then further refined to its ultimate degree of purity by adsorption.

PURITY GUARDIANSHIP

Since white oil was introduced initially as a medicine it followed logically that national pharmacopoeias assumed responsibility for establishing quality and purity criteria for white oils. In the United States, this responsibility was taken by the *U.S. Pharmacopoeia* and the *National Formulary*, whose legal status is based on the adoption of its definitions and standards in the Food, Drug, and Cosmetic Act. In the United Kingdom, the *British Pharmacopoeia* became responsible for establishing, maintaining, and strengthening the rigid standards for white mineral oil purity. The requirements of the two pharmacopoeias are substantially the same. The prime objective of specifications for pharmaceutical oils is to ensure the product is as inert as possible and free of toxic materials. Differences in the key specifications for white oils among the two pharmacopoeias and the *National Formulary* are shown in Table 1.

The *U.S. Pharmacopoeia* specification includes an ultraviolet light absorbance test as a requirement for mineral oil which is essentially the same as ASTM Evaluation of White Mineral Oils by Ultraviolet Absorption (D 2269). The procedure measures the absorbance of a dimethyl sulfoxide extract.

Since, during manufacture, natural inhibitors in petroleum oils are removed, the resultant white oils are not stable to oxidation. Both the *U.S. Pharmacopoeia* and the *National Formulary* permit the addition of a suitable stabilizer. The *British Pharmacopoeia* permits the addition of tocopherol or butylated hydroxytoluene in a proportion not greater than 10 ppm.

Besides pharmaceutical uses, the nonreactivity of a white oil makes it suitable: (1) as a carrier for alkali dispersions for use in organic chemical reactions; (2) in textile and precision instrument lubricants where nongumming is desired; (3) as a tobacco desuckering oil; and (4) in many human, plant, and animal preparations where nontoxicity and freedom from irritation are desired. Each of these uses may require unique qualities.

ASSESSMENT OF QUALITY

Many test methods are available to determine whether the quality for white oils exists as prescribed by the pharmacopoeias, the *National Formulary*, and other consumers. The following paragraphs will outline the significance of the most important of these tests.

Acid Test (Readily Carbonizable Substances)

The acid test, ASTM Test for Carbonizable Substances in White Mineral Oil (Liquid Petrolatum) (D 565), is a measure of the pu-

TABLE 1. Pharmaceutical grades of white oils.

	Mineral Oil, USP XXI, 1985	Light Mineral Oil, NF XVI, 1985	Liquid Paraffin, BP 1980
Description			transparent, colorless, oily liquid free from fluorescence in daylight; almost odorless and tasteless
Specific gravity, 25°C	0.845 to 0.905	0.818 to 0.880	
Viscosity, cSt	34.5 min at 40°C	33.5 max at 40°C	64 min at 37.8°C
Solid paraffins		black line (0.5 mm) visible through a 25-mm layer at 0°C	
Acidity/alkalinity		nil	
Sulfur compounds		negative to sodium plumbite test	
Carbonizable substances		maximum color of acid layer stipulated	
Polynuclear compound Absorbance			
260 to 350 nm[a]	0.1 max	0.1 max	
A 1 cm 240 to 280 nm, 2.0% solution			0.10 max

[a]On dimethyl sulfoxide extract.

rity of a white oil and its freedom from aromatic, unsaturated, and other materials reactable with sulfuric acid.

In the test, equal portions of oil and 94.7 ± 0.2 percent sulfuric acid are shaken in a prescribed manner at 100°C (212°F) for 10 min, and the color of the acid layer is compared with that of a standard color solution. The more fully refined the oil, the lighter the color of the acid layer.

Most white oils manufacturers, for their own information, protection, and process control, have gone beyond this pharmacopoeia requirement and developed rating systems to measure the degree of color of the acid layer and thus assure the highest levels of the oil purity.

Aniline Point

The ASTM Test for Aniline Point and Mixed Aniline Point of Petroleum Products and Hydrocarbon Solvents (D 611/IP 2) provides a measure of the aromaticity and naphthenicity of an oil. In the test, equal volumes of aniline and oil are mixed and heated until a miscible mixture is formed. On cooling at a prescribed rate, the temperature at which the mixture becomes cloudy is recorded and identified as the aniline point. While not typically applicable to pharmaceutical grade white oils, it is employed in technical oil specifications as a measure of degree of refinement and type base oil stock. For any particular oil fraction, a higher degree of refinement is reflected by an increase in aniline point. Aniline point also increases with the average molecular weight of the oil as well as with increasing proportions of paraffinic hydrocarbons to naphthenic hydrocarbons. Aniline point specifications also can be used to advantage for light technical grade white oils used in agricultural sprays where the presence of aromatic hydrocarbons might cause foliage damage.

Cloud Point

ASTM Test for Cloud Point of Petroleum Oils (D 2500/IP 219) provides the temperature at which wax appears in an oil. This information is significant for oils to be used at low temperatures where precipitation of wax might affect the performance of the oil. Examples include refrigerator and hydraulic oils.

The cloud point test is considered to be more meaningful than the "Solid Paraffins" test called for by the pharmacopoeias which determines whether or not wax precipitates out at 0°C (32°F).

Color and Transparency

By definition, a pharmaceutical grade white oil must be colorless and transparent. On the Saybolt scale, water has a value of +25 while typical white oils have color ratings of +30. In other words, pharmaceutical grade white oils have less color than water.

Freedom from color and transparency are the unique properties of white oils that make them useful in their many cosmetic, pharmaceutical, agricultural, food, and industrial applications. Among the latter are textiles, plastics, precision instruments, leather, household specialties, and printing ink.

White oil color is determined by means of ASTM Test for Saybolt Color of Petroleum Products (Saybolt Chromometer Method) (D 156). In this instrument, the height of a column of the oil is decreased by levels corresponding to color numbers until the color of the sample is clearly lighter than that of the standard. The color number immediately above this level is recorded as the Saybolt color of the oil.

Dielectric Breakdown Voltage

Dielectric breakdown voltage is the measure of the ability of an oil to withstand electrical stress. White oils used in the electrical or electronics industries as a coolant, insulating fluid, lubricant, or cleaner must have a high-dielectric strength. Dielectric breakdown voltage is determined by ASTM Test for Dielectric Breakdown Voltage of Insulating Liquids Using Disk Electrodes (D 877).

Distillation Range

The distillation range for very white light oils provides information on volatility, evaporation rates, and residue remaining after evaporation. Such data is important for agricultural and household sprays, agricultural product processing, and printing inks. The baking and plastic industries often include initial boiling point temperature or the minimum allowable temperature at which the first several percent of the oil comes overhead during distillation or both as part of their specifications for white oils.

The distillation range on low-viscosity grades of mineral oils is determined by ASTM Test for Distillation of Petroleum Products (D 86/IP 123). Distillation range for higher boiling grades is measured by ASTM Test for Distillation of Petroleum Products at a Reduced Pressure (D 1160).

Flash and Fire Point

Flash and fire points of white oils are a measure of their fire hazard. High-flash and fire points are particularly important in many industrial applications where the oils are used at high temperatures.

Flash and fire points are determined by ASTM Test for Flash Point by Pensky-Martens Closed Tester (D 93/IP 34) or by ASTM Test for Flash and Fire Points by Cleveland Open Cup (D 92/IP 36). The Cleveland open cup method is also used to determine fire point.

Both methods measure flash point at the minimum temperature to which the oil must be heated to provide a sufficient amount of vapor to ignite. Since the Pensky-Martens tester is a closed system, flash point values are lower with it than those determined by the Cleveland method.

The fire point is the temperature at which the oil ignites and burns for 5 s.

Fluorescence

The *British Pharmacopoeia* states that the oil shall be free from fluorescence by daylight. Fluorescence is a measure of purity and indicates the presence of aromatic hydrocarbons.

Examination of a sample bottle of oil under ultraviolet light is an even more sensitive test which many manufacturers employ as a rapid means of detecting incompleteness of refining or the presence of petroleum contamination. While it is possible to obtain fluorescent oils that are free of polynuclear aromatic compounds, oils that do not fluoresce are virtually certain to be free of such compounds.

Neutrality

In the *U.S. Pharmacopoeia,* the test for neutrality is 10 ml of oil boiled with 10 mL of ethyl alcohol, and the alcohol is tested by litmus paper. The *British Pharmacopoeia* employs 10 mL of alcohol, using 5 g of oil.

This is an extremely sensitive test in that the presence of as little as 0.01 mg potassium hydroxide/g is sufficient to turn red litmus paper blue. The test is largely a relic of earlier times when it was thought that detectable amounts of chemicals used in processing could remain in the finished product.

Odor and Taste

The *British Pharmacopoeia* requires that white oil be almost odorless and almost tasteless. Both tests are highly subjective, but, with practice, many inspectors can detect the slightest trace of an off odor or taste.

Peroxide Value

The peroxide value is an accelerated test to measure the oxidation stability of an oil. Although stability is not an expressed requirement of the pharmacopoeias, sufficient stability is needed to ensure adequate shelf life without the development of rancidity, acidity, or color degradation.

Many consumers of white oils in the pharmaceutical, cosmetic, textile, and coatings industries prescribe oxidation stability tests in their purchase specifications. When white oils are used as heating fluid baths or for other uses at elevated temperatures where medicinal purity is not prescribed, more stabilizer may be used to increase stability.

Pour Point

Pour point provides a means of determining the type petroleum feedstock from which the white oil was manufactured or its previous processing history. It also reflects the presence of wax or paraffinic hydrocarbons. Clearly, in any application where the white oil is used at low temperatures or the oil is subjected during handling or storage to low temperatures, the pour point is important and, in many industrial applications, critical.

In the ASTM Test for Pour Point of Petroleum Oils (D 97/IP 15), the oil is heated to a specified temperature which is dependent on the anticipated pour-point range, cooled at a specified rate, and examined at 3°C (5°F) intervals for flow. The lowest temperature at which no movement of the oil is detected is recorded. The 3°C (5°F) temperature value immediately preceding the recorded temperature is defined as the pour point.

Refractive Index

The refractive index is the ratio of the velocity of light in air to its velocity in the substance under examination. It is used, together with density and viscosity measurements, in calculating the paraffin-naphthene ratio in white oils. Because refractive index is a measure of aromaticity and unsaturation on a given stock, manufacturers also use it as a means of process control.

Refractive index is measured using ASTM Test for Refractive Index and Refractive Dispersion of Hydrocarbon Liquids (D 1218).

Smoke Point

Smoke point is of particular interest to industries, such as the baking industry, whose processes expose or use the oil at extremely high temperatures.

The smoke-point test is conducted in a black box suitably vented and illuminated to permit detection of white vapors. The oil is carefully heated under specified conditions until the first consistent appearance of vapors is detected. The temperature of the oil at that time is recorded as the smoke point.

Gravity, Density and API Gravity

Density, specific gravity, or American Petroleum Institute (API) gravity may be determined by ASTM Test for Density, Relative Density (Specific Gravity), or API Gravity of Crude Petroleum and Liquid Petroleum Products by Hydrometer Method (D 1298/IP 160). The terms are defined as follows:

Specific Gravity—The ratio of the mass of a given volume of liquid at 15°C (60°F) to the mass of an equal volume of pure water at the same temperature.

Density—The mass (weight in vacuo) of liquid per unit volume at 15°C as stated for the specified hydrometer.

API gravity is a special function of specific gravity that was arbitrarily established to permit gravity calculations in whole numbers. It is related to specific gravity by the following formula:

$$\text{API gravity, deg} = \frac{141.5}{\text{sp gr at } 15/15°\text{C } (60/60°\text{F})} - 131.5$$

Density, specific gravity, and API gravity values permit conversion of volumes at the measured temperature to volumes at the standard petroleum temperatures of 15°C and 60°F. Calculation to weight is possible where compositions are formulated on a weight basis. At a given viscosity density, specific gravity and API gravity provide a means for determining whether a white oil is derived from a paraffinic or naphthenic stock.

Sulfur Compounds

The total sulfur content of white oils normally is well below 100 ppm because of the severe refining to which the oil has been subjected. Therefore, sulfur is not a limiting factor in the specifications of the *U.S. Pharmacopoeia* or the *National Formulary.*

The *U.S. Pharmacopoeia* XVII carries a sulfur compounds test which is similar to the ASTM doctor test, although less sensitive. This test is required by the Food and Drug Administration on all food grade white mineral oil.

Unsulfonated Residue

ASTM Test for Unsulfonated Residue of Petroleum Plant Spray Oils (D 483) provides a crude measure of the reactable hydrocarbons present in an oil. It is often included in specifications for oils intended for use as foliage sprays where aromatic and unsaturated hydrocarbons would cause foliage damage. For high-quality white oils the readily carbonizable substance test is used because it is a much more sensitive measure for such impurities.

Ultraviolet Absorption Tests

Considerable concern has been generated in recent years over the possible presence of carcinogenic polynuclear aromatic hydrocarbons in white oils. Because of this concern much effort has been expended to develop tests to detect the smallest traces of these carcinogens.

ASTM Test D 2269 measures the absorbance over the wavelength range of 260 to 350 nm in a 10 mm cell of a dimethyl sulfonide extract of the oil. The polynuclear aromatic hydrocarbons present in the white oils are concentrated. The Food and Drug Administration has recognized that there is little likelihood of carcinogenic hydrocarbons being present if the absorbance of the extract in the wavelength range is no greater than 0.1. This level of absorbance is roughly equivalent to 5 ppm polynuclear hydrocarbons with a test sensitivity of 0.3 ppm.

Viscosity

Viscosity is one of the most important properties to be considered in the evaluation of a white oil. From a functional point of view, requirements for viscosity vary widely according to the user for which the oil is intended and may be as low as 4 cSt or as high as 70 cSt. White oil for internal use generally should have high viscosity in order to minimize possibilities of leakage. No general statement can be made concerning the viscosity of white oil for other uses.

ASTM Test Method for Kinematic Viscosity of Transparent and Opaque Liquids (and the Calculation of Dynamic Viscosity) (D 445) (IP 71) is used for the determination of viscosity of white oils.

Wax Precipitation Point

The wax precipitation point test is unique to the refrigerator and freezer industries where oil is added to the Freon refrigerant to provide lubrication during the expansion and compression cycles. Just prior to expansion the refrigerant bearing the oil passes through a length of capillary copper tubing. It is critical to the process that no wax precipitates and clogs the capillary tubing.

Applicable ASTM/IP Standards

ASTM	*IP*	*Title*
D 86	123	Distillation of Petroleum Products
D 92	36	Flash and Fire Points by Cleveland Open Cup
D 93	34	Flash Point by Pensky-Martens Closed Tester
D 97	15	Pour Point of Petroleum Oils
D 156		Saybolt Color of Petroleum Products
D 287		API Gravity of Crude Petroleum and Petroleum Products (Hydrometer Method)
D 445	71	Kinematic Viscosity of Transparent and Opaque Liquids
D 447		Distillation of Plant Spray Oils
D 483		Unsulfonated Residue of Petroleum Plant Spray Oils
D 565		Carbonizable Substances in White Mineral Oil (Liquid Petrolatum)
D 611	2	Aniline Point and Mixed Aniline Point of Petroleum Products and Hydrocarbon Solvents

Applicable ASTM/IP Standards *(continued)*

ASTM	*IP*	*Title*
D 877		Dielectric Breakdown Voltage of Insulating Liquids using Disk Electrodes
D 1160		Distillation of Petroleum Products at Reduced Pressure
D 1218		Refractive Index and Refractive Dispersion of Hydrocarbon Liquids
D 1298	160	Density, Relative Density (Specific Gravity), or API Gravity of Crude Petroleum and Liquid Petroleum Products by Hydrometer Method
D 2269		Evaluation of White Mineral Oils by Ultraviolet Absorption
D 2500	219	Cloud Point of Petroleum Oils

Bibliography

British Pharmacopoeia, 1980.

Dunstan, A. E., Nash, A. W., et al, *The Science of Petroleum, Vol. IV*, Oxford University Press, London, England, 1938.

Franks, A. J., "New Technology of White Mineral Oils," *American Perfumer*, Vol. 76, No. 2, 1961.

Meyer, E., *White Mineral Oil, Petrolatum and Related Products*, Chemical Publishing Company, New York, 1968.

"Mineral Oil," 21 CFR 178.3620.

"Oderless Light Petroleum Hydrocarbons," 21 CFR 172.884.

"Petroleum Wax," 21 CFR 172.886.

The National Formulary XVI, 1985.

United States Pharmacopoeia XXI, 1985.

"White Mineral Oil," 21 CFR 172.878.

12

Petroleum Oils for Rubber

INTRODUCTION

Historical Background

PETROLEUM PRODUCTS covering a molecular weight range that includes light solvents, lubricating oils, waxes, and residual materials have been used with rubber for at least 150 years. Today, these same products are still used, and, since the advent of synthetic rubber, their use has reached greater proportions than ever before. This chapter, however, will be limited to petroleum oils.

Why are such products usable when one normally thinks of oils as the enemy of rubber? The explanation is that the degrading effect of oils on vulcanized rubber is put to use as a softener for unvulcanized rubber. Rubber, both natural and synthetic, is elastic in the raw state. It must be broken down by mechanical means to a more plastic state prior to adding the necessary compounding ingredients. Oils are added to accelerate the masticating effect which makes the rubber more workable and pliable. Materials that cause this softening effect are called softeners. They also are called processing aids or plasticizers, although the latter term is usually reserved for plastics.

Natural Rubber

In the early days of the rubber industry, natural rubber was the only rubber polymer used. It is characteristic of natural rubber to break down fairly easily when milled, and, as a consequence, little softener is required. While petroleum oils are usable for this purpose, it was found that naval stores derivatives, such as pine tar, not only aided breakdown but also facilitated the development of tack in the mixture. Tack is the ability to cohere, and it is a highly desirable property for certain rubber applications. For this reason, pine tar became a preferred softener in all natural rubber compounds, and it is used to some extent today. In current practice, however, natural rubber is used frequently in blends with synthetic rubber. Because petroleum oils are less expensive and work well with this combination, they now are used generally for plasticization.

Synthetic Rubber

The preparation of a synthetic rubber equivalent to natural rubber has long been a research laboratory goal. This has been particularly true in countries vulnerable to having their rubber supply shut off in time of war. With this motivation, both Germany and the United States devoted much effort to the problem prior to World War II. This culminated in the development of synthetic rubber in both countries, with some of the German technology licensed in the United States.

Synthetic rubber reached large-scale production during World War II. Although it was a reasonable substitute, it was not a duplicate for natural rubber. One place where the lack of similarity between natural and synthetic rubber showed up was in factory processing. Synthetic rubber proved to be much more difficult to process. This problem offered a real technical challenge to the rubber compounder when synthetic rubber was first commercialized. While many different schemes to overcome this deficiency were tried, the use of softeners, plus better quality control in the polymer plants, proved to be a reasonably satisfactory solution.

Early efforts focused on the hunt for the best softener to use with the butadiene-styrene polymer (SBR) which was chosen for commercialization in this country.

Based on performance, cost, and availability, petroleum oils proved to be the best of several alternatives.

During the investigations on the use of oils, it was found that more oil was required for synthetic rubber than for natural rubber. Thus, oils and their properties became more important to the compounder both from the processing standpoint and the effect on physical properties of the final rubber product. From the standpoint of the supplier, the increased demand for rubber oils made them an important specialty product.

Oil-Extended Rubber

At the close of World War II, the synthetic rubber plants in production were adequate to supply United States' needs. With the advent of the Korean War, a shortage was foreseen—particularly since a "guns and butter" economy existed. An ingenious technical innovation averted any shortage and resulted in a modified synthetic rubber with properties which were highly satisfactory for many applications.

This innovation carried the degree of monomer polymerization to a higher level than was done normally, which was then cut back or extended with substantial quantities of oil. The cut back was necessary to get the rubber back to a workable level since the higher degree of polymerization produced a polymer that was impossible to process by itself. This technique stretched the SBR supply of the country without a corresponding need to expand styrene and butadiene monomers production. Today, both extended and nonextended synthetic rubbers are produced. In the case of SBR, production of extended exceeds nonextended.

In terms of quantities of oil used, about 5 to 10 percent oil is added to nonextended rubber. In the oil extended polymers the amount of oil may represent 35 percent or more of the overall polymer.

COMPOSITION OF RUBBER OILS

Petroleum oils for use in rubber are generally in the lubricating oil viscosity range, although residual materials are sometimes employed. The importance of composition of these oils lies in the effect composition has on their compatibility with rubber.

Rubber-plasticizer systems may be thought of as solutions in which plasticizing oils are not used in sufficient quantity to form a liquid external phase. As with all such systems, solubility is dictated by the nature of the constituents. In the case of rubber-oil mixtures, insolubility is manifested by "sweatout" or exudation of oil to the surface of a rubber article—usually most pronounced after vulcanization. To avoid the sweatout problem and benefit the rubber compound in a variety of ways, good compatibility is desirable to the rubber compounder. Benefits include lowered volatility of oil from polymer, faster incorporation of compounding ingredients during mixing, better dispersion of reinforcing pigments which correspondingly aids physical properties, reduction of oil takeup time and heat generation, and numerous other effects.

As the interest in processing oils and their usage grew, composition assumed more importance. Two basic methods for obtaining compositional information about rubber oils came into use—molecular and carbon type analysis.

Molecular Type Analysis

Molecular type analysis separates an oil into different molecular species. A molecular type analysis is the so-called "clay-gel analysis." In this method, group separation is achieved by adsorption in a percolation column with selected grades of clay and silica gel as the adsorption media. This procedure is now standardized as ASTM Test for Characteristic Groups in Rubber Extender and Processing Oils by the Clay-Gel Adsorption Chromatographic Method (D 2007).

Carbon Type Analysis

Although informative, molecular type analytical methods are cumbersome and time consuming. For these reasons, there has been great interest in the correlation of composition with physical properties. Among the most definitive of these efforts was the refractive index-density-molecular weight (n-d-M) method reported by Van Nes and Van Westen. Van Nes and Van Westen demonstrated the n-d-M method is a reliable procedure for determining carbon type composition of oil fractions. The method has been standardized as ASTM Test for Carbon Distribution and Structural Group Analysis of Petroleum Oils by the n-d-M Method (D 3238). Carbon type composition gives the breakdown of total carbon atoms between various structures.

Kurtz and associates developed a correlative method for determining carbon type distribution from viscosity-gravity constant (VGC) and refractivity intercept (r_i). Both these functions are independent of molecular weight which makes them well suited for characterizing oils of widely different viscosities. These workers showed that it is possible to superimpose lines of constant VGC and r_i on a triangular graph of carbon type composition. Then, if VGC and r_i are calculated for an oil whose composition is desired from viscosity, gravity, and refractive index, a point is fixed on the graph which uniquely establishes the percentage of aromatic ring structures (C_A), naphthene ring structures (C_N), and diolefin structure (C_D) for the oil. This method has been standardized as ASTM Test for Carbon-Type Composition of Insulating Oils of Petroleum Origin (D 2140).

Sweely and co-workers were also the first to apply physical property methods for composition to rubber process oils. Since the physical properties required are easily obtainable, they present a convenient way for the rubber technologist whose laboratory is not set up for analysis of petroleum oils to obtain compositional information on process oils. Usually physical property methods are applied to the analysis of the whole oil; however, if more detailed information is required, the oil can be given a molecular type separation into an aromatics and a saturates fraction and the carbon type composition of each determined.

As a summary for this section on composition, analysis based on molecular or carbon type distribution provides the rubber compounder with sound information. For research work, instrumental methods such as high-resolution mass spectrometry, nuclear magnetic resonance, and infrared spectroscopy are being used more frequently. In addition, new analytical techniques for polar compounds are much faster than conventional methods.

IMPORTANCE OF COMPOSITION TO COMPOUNDERS

What does the determination of oil composition mean to the rubber compounder? First of all, both molecular and carbon type compositions give a measure of aromaticity of oils. The aromatic content of an oil determines its compatibility with many rubbers. Thus, aromatic content dictates whether an oil is acceptable for use in a given rubber compound or not.

The degree of aromaticity affects performance of an oil in elastomers, particularly with the more polar types such as butadiene-acrylonitrile rubbers and polyvinyl chloride. In this case, carbon type analysis can be more informative than molecular type.

Asphaltenes present in an oil are determined by pentane precipitation. They indicate the presence of resinous oxidation products or residual material which may be undesirable to the rubber compounder since such products often lack batch-to-batch uniformity.

All petroleum oils contain oxygen, nitrogen, and sulfur compounds to some degree. If these nonhydrocarbons or polar compounds are reactive, as would be the case with acidic or basic compounds, they could impart an adverse effect on the vulcanization characteristics of the rubber with which they are used. They may also cause discoloration and staining of light colored rubber goods under exposure to sunlight or ultraviolet light. For these reasons, oils containing high contents of polar compounds usually are checked for any adverse effects before acceptance for general usage.

CLASSIFICATION OF RUBBER OILS

To aid the user of rubber process oils, a classification system has been developed based on some of the properties just discussed. This classification, which is given in ASTM Classification for Various Types of Petroleum Oils for Rubber Compounding Use (D 2226) is shown in Table 1.

Table 1 does not list aromatics; however, an indirect measure of aromaticity is given the saturates limits. Saturates are used instead of aromatics for classification purposes because aromatics must be desorbed from adsorbent in the clay-gel method, whereas the saturates are determined directly after removal of solvent.

Observe also in Tabel 1 that VGC is used to differentiate between Type 104A and 104B oils. In this case, VGC is being used as an estimator of aromaticity. In addition to its use for determining carbon type composition in the viscosity-gravity constant/refractivity intercept method, VGC by itself has long been used to indicate aromaticity of rubber oils. VGC is a function independent of molecular weight and quite useful for comparing different oils. Viscosity-gravity constant has been standardized as ASTM Calculation of Vicosity-Gravity Constant (VGC) of Petroleum Oils (D 2501).

One other measure of aromaticity that has received a considerable degree of acceptance in the rubber industry is ASTM Test for Aniline Point and Mixed Aniline Point of Petroleum Products and Hydrocarbon Solvents (D 611). Unfortunately, aniline point is not molecular weight independent. This results in oils with the same aromatic contents, but different viscosities that have different aniline points. Other methods should be used if a more precise measure of aromaticity is required.

PHYSICAL PROPERTIES OF RUBBER OILS

In general, rubber compounders use tests to control uniform quality of rubber oils and predict how the oils will affect the rubber compound or oil-extended rubber. Standard physical property tests used by the petroleum industry satisfy these requirements.

The following paragraphs describe the significance of physical property tests that normally are used for rubber oils. Details on individual tests can be found in the appropriate *Annual Book of ASTM Standards.*

Specific Gravity

Specific gravity provides a means to convert volumes to weights and vice versa. Since rubber compounding is done on a weight basis, a considerable volume effect in rubber can be obtained with oils of different gravities. Knowledge of specific gravity permits the calculation of the magnitude of the difference.

For oils of similar viscosity, specific gravity can be used as a rough measure of aromaticity (the higher the specific gravity, the greater the aromaticity).

As described in earlier paragraphs, specific gravity also is needed to calculate VGC and refractivity intercept in the determination of carbon type composition.

ASTM Test for Density, Relative Density (Specific Gravity), or API Gravity of Crude Petroleum and Liquid Petroleum Products by Hydrometer Method (D 1298) is used to determine specific gravity.

TABLE 1. Classification of oil types.

Types	Asphaltenes, max %	Polar Compounds, max %	Saturated Hydrocarbons, %
101	0.75	25	20 max
102	0.5	12	20.1 to 35
103	0.3	6	35.1 to 65
104[a]	0.1	1	65 min

[a]Type 104 oils are further classified into two subtypes: 104A and 104B for SBR polymers only. Type 104b oils have a viscosity-gravity constant of 0.820 max; type 104A have a viscosity-gravity constant greater than 0.820.

Viscosity

Viscosity gives a measure of the flow properties of the oil which, in turn, determines the ease of handling at various temperatures. Viscosities of oils at low temperature have been shown to correlate with low-temperature properties of rubber containing these oils. Because viscosity is a measure of molecular weight, it can be used to estimate the compatibility of oils in polymers. Polymers with a critical tolerance for oil will not hold large volumes of oil of high-molecular weight. Viscosity for rubber oils is measured by ASTM Test for Kinematic Viscosity of Transparent and Opaque Liquids (and the Calculation of Dynamic Viscosity) (D 445).

Color

Oil color is an indicator of rubber oil suitability for use in light color rubber compounds. It also influences polymer color in oil-extended polymers. Initial oil color cannot be used as a measure of color stability of an oil itself or of a rubber compound containing it. Color is determined by ASTM Color of Petroleum Products (ASTM Color Scale) (D 1500).

Ultraviolet Absorptivity

For oils of a similar type, ultraviolet (UV) absorptivity is a good indicator of the resistance of an oil to discoloration under exposure to artificial or natural light. Oils with low absorptivities at 260 nm have been found to impart good color stability to light-colored rubber compounds. ASTM Test for Ultraviolet Absorbance and Absorptivity of Petroleum Products (D 2008) is used to measure UV absorptivity.

Refractive Index

ASTM Test for Refractive Index and Refractive Dispersion of Hydrocarbon Liquids (D 1218) is a convenient test for establishing batch-to-batch continuity in rubber oil shipments. Refractive index is needed also to calculate the refractivity intercept in the determination of carbon type composition.

Aniline Point

ASTM D 611 gives a rough measure of aromaticity of an oil which is useful in the prediction of swelling characteristics of rubbers exposed to the oil. This test should be used with caution since aniline point is molecular weight dependent.

Flash Point

Rubber compounders often use flash point as a measure of oil volatility. Volatility is important because rubber products are exposed to elevated temperatures during mixing operations and, oftentimes, in service. Although flash point has a certain utility for this purpose, it gives no indication of the amount of low-boiling material present. Therefore, a distillation curve should be used when volatility is a critical factor.

Flash point also is used to determine fire hazard aspects, particularly when use of oils with low-flash points is being considered. Flash point is determined by ASTM Test for Flash and Fire Points by Cleveland Open Cup (D 92/IP 36) or by ASTM Test for Flash Point by Pensky-Martens Closed Tester (D 93/IP 34).

Evaporation Loss

ASTM Test for Evaporation Loss of Lubricating Grease and Oils (D 972) gives a measure of oil volatility under controlled conditions and is used frequently for specification purposes. However, because volatility of oil from a rubber compound may be influenced by its compatibility with the rubber, a volatility test of the compound often is made under laboratory test conditions pertinent to the intended service.

Pour and Cloud Point

Pour and cloud point tests are useful in establishing low-temperature characteristics of an oil. Pour point is mainly of concern in connection with handling oils at low temperatures. Cloud point is useful in showing whether any wax is present in an oil. ASTM Test for Pour Point of Petroleum Oils (D 97) normally is used to determine pour point of rubber oils.

SPECIALTY APPLICATIONS OF PROCESS OILS

While rubber affords a substantial outlet for processing oils, it is by no means the only outlet for materials of this type. Other areas (confined to polymeric materials) where mineral oils can be used include caulks and sealants, textile specialties, resin extenders, polyvinyl chloride, and other polymer plasticizers and adhesives.

In these product areas, a wide variety of specialized oils are used. Requirements range from high-priced oils that satisfy Food and Drug Administration regulations to the cheapest oil available where product cost is the only consideration. Process oil producers are sufficiently versatile to meet any reasonable requirements. ASTM test methods and specifications simplify the task.

Applicable ASTM/IP Standards

ASTM	*IP*	*Title*
D 92	36	Flash and Fire Points by Cleveland Open Cup
D 97	15	Pour Point of Petroleum Oils
D 445	71	Kinematic Viscosity of Transparent and Opaque Liquids (and the Calculation of Dynamic Viscosity)
D 611		Aniline Point and Mixed Aniline Point of Petroleum Products and Hydrocarbon Solvents
D 972		Evaporation Loss of Lubricating Greases and Oils
D 1218		Refractive Index and Refractive Dispersion of Hydrocarbon Liquids
D 1298	160	Density, Relative Density (Specific Gravity), or API Gravity of Crude Petroleum and Liquid Petroleum Products by Hydrometer Method
D 1500	196	ASTM Color of Petroleum Products (ASTM Color Scale)
D 2007		Characteristic Groups in Rubber Extender and Processing Oils by the Clay-Gel Adsorption Chromatographic Method
D 2008		Ultraviolet Absorbance and Absorptivity of Petroleum Products
D 2140		Carbon-Type Composition of Insulating Oils of Petroleum Origin
D 2226		Classification for Various Types of Petroleum Oils for Rubber Compounding use
D 2501		Viscosity-Gravity Constant (VGC) of Petroleum Oils